Einfuhrung

die Kosmischen Rätsel der Puppenspieler

This book is about The Puppeteer's Cosmic Puzzle, a novel deck of 48 playing cards related to the standard deck of 52 in use throughout the world today. Suits in the Puzzle deck graphically answer the 6 little questions how many, why, who, what, when, and where.

Dieses Buch ist über den Kosmische Puzzle des Puppeteer, ein neues Satz von 48 Spielkarten mit einer Beziehung zu dem Standardsatz von 52 im Einsatz auf der ganzen Welt heute. Farben in dem Puzzle-Satz grafisch beantworten Sie die 6 kleinen Fragen, wie viele, warum, wer, was, wann und wo.

The Puppeteer represents nature at work, and the natural order. With that, plus some abbreviation, approximation, and time-honored convention, a playing-card microcosm can be pieced together. Ancient symbols for the planets, the zodiac, and others help to indicate cosmic phenomena that can be verified by naked eye observations.

Die Puppeteer stellt die Natur bei der Arbeit, und die natürliche Ordnung. Damit plus einige Abkürzung, Annäherung und altehrwürdige Konvention, ein Spielkartenmikrokosmos zusammen genäht werden. Alte Symbole für die Planeten, der Tierkreis, und andere helfen kosmische Phänomene zeigen, die mit bloßem Auge Beobachtungen überprüft werden können.

Big Answers to Little Questions is a stand-alone activity and coloring book that includes some card games. All the cards are pictured, named, and explained. The front cover shows the Puppeteer surrounded by the 36 cards in the 6 suits mentioned. The back cover shows the Puppeteer's Eye in Hand surrounded by 2 more suits that depict the 12 constellations of the ecliptic.

Gross Antworten auf kleine Fragen ist ein Stand-alone-Aktivität und Malbuch, das einige Kartenspiele umfasst. Alle Karten sind im Bild, mit dem Namen, und erläutert. Die Frontabdeckung zeigt der Puppenspieler umgeben von 36 Karten in den Farben 6 erwähnt. Die hintere Abdeckung zeigt die Puppeteer Auge in der Hand, umgeben von 2 mehr Farben, die die 12 Konstellationen der Ekliptik darstellen.

This book and the unique playing card puzzle that it's about are archived in the Playing Card Museum of France in Issy-les-Moulineaux.

Dieses Buch und das einzigartige Spielkarte Puzzle, dass es über ist in dem Spielkartenmuseum von Frankreich in Issy-les-Moulineaux archiviert.

Inhaltsverzeichnis

Farben

Spiele

Matrix

不是

Solitar

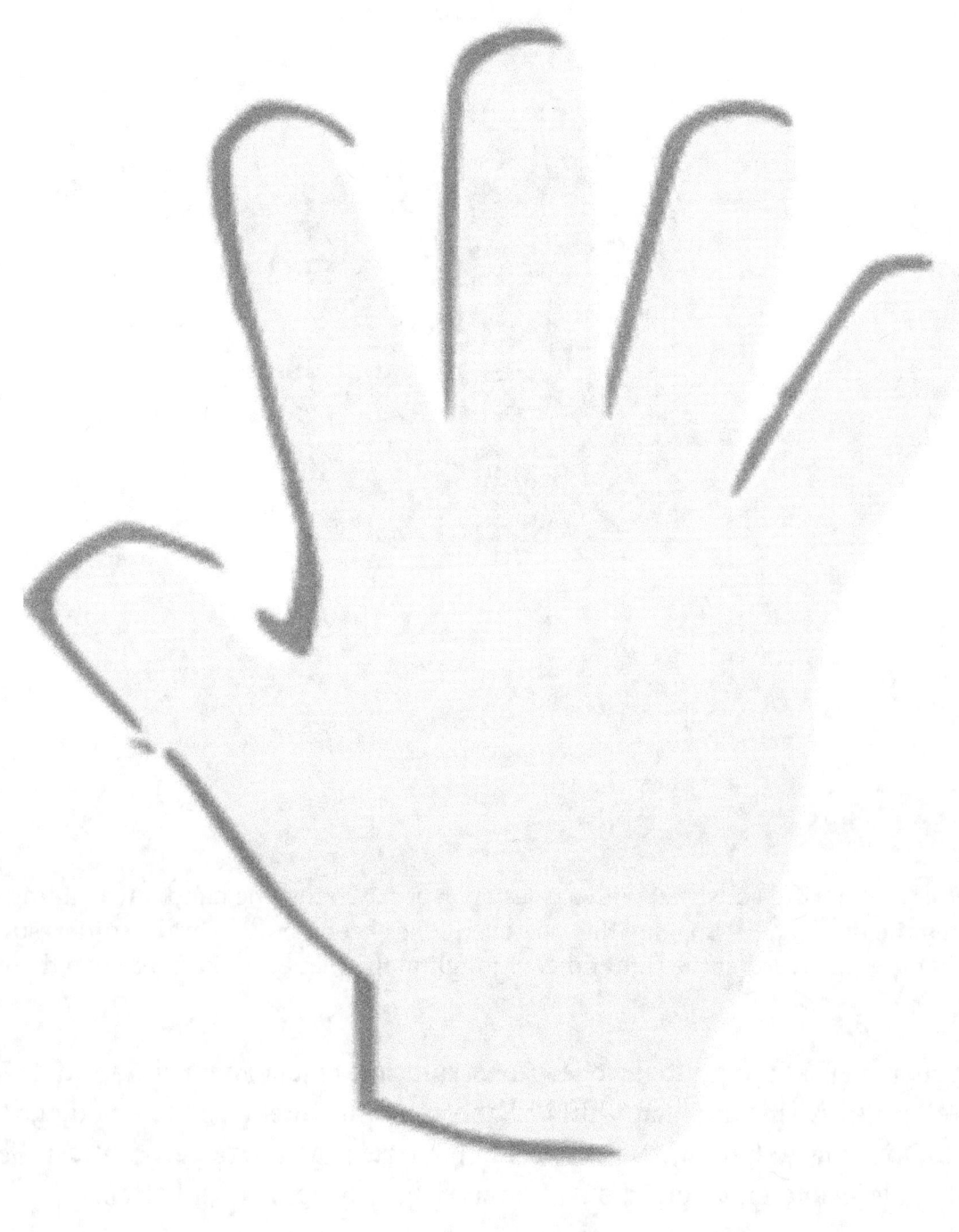

Wie Viele?

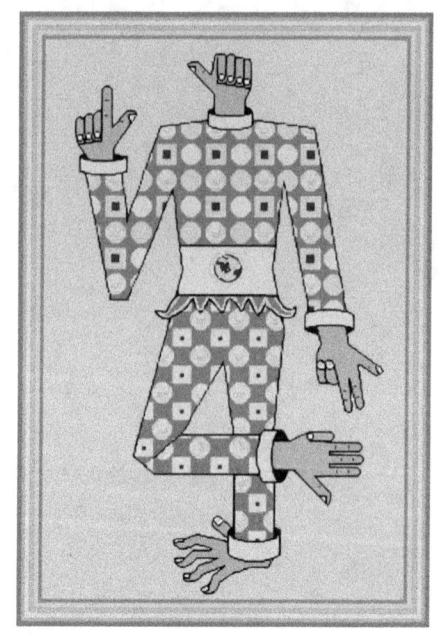

Die Hand des Puppeteer

The *Puppeteer* is a fanciful construct with a celestial aspect. On the one hand, great at finger counting. On the other hand, imagine this one-eyed Joker keeping the world in order, such as it is, by playing the right card at the right time, or juggling the planets to keep them in their proper orbits.

Der Puppenspieler ist ein fantasievolles Konstrukt mit einem Himmel Aspekte. Auf der einen Seite, groß bei Fingerzählen. Auf der anderen Seite, stellen diese einäugigen Joker die Welt in Ordnung zu halten, wie es ist, durch die richtige Karte zur richtigen Zeit zu spielen, oder die Planeten jongliert sie in ihren richtigen Bahnen zu halten.

Or simply imagine this extraordinary puppeteer playing with puppets; with shoes and a life sized puppet head on, or working several little hand puppets at once, putting on quite a show.

Oder einfach vorstellen, diese außergewöhnliche puppeteer spielt mit Puppen; mit Schuhen und einer lebensgroßen Puppe Kopf oder mehrere kleine Handpuppen auf einmal arbeiten, auf ganz eine Show.

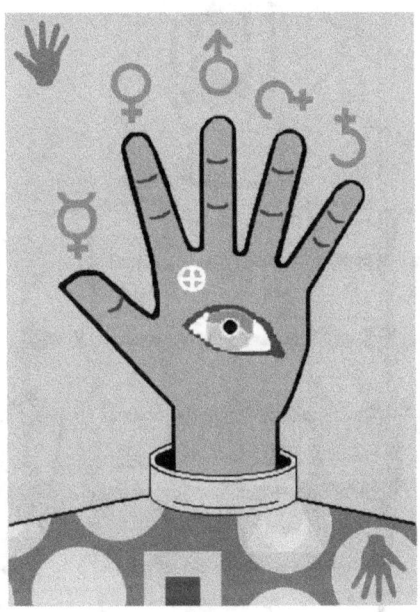

The *Eye in Hand* represents the earth, moon, and sun surrounded by the five planets visible to the naked eye. Fingers represent the planets in order of their apparent speed relative to the fixed stars, fastest to slowest left to right: thumb/1st/Mercury, index/2nd/Venus, middle/3rd/Mars, ring/4th/Jupiter, pinky/5th/Saturn.

Das Auge in der Hand steht für die Erde, Mond und die fünf Planeten mit bloßem Auge umgeben Sonne. Finger stellen die Planeten in der Reihenfolge ihrer scheinbaren Geschwindigkeit relativ zu den Fixsternen, schnellsten zum langsamsten links nach rechts: Daumen / 1. / Mercury, Index / 2. / Venus, Mitte / 3. / Mars, Ring / 4. / Jupiter, Pinky / 5. / Saturn.

The eye in the hand is the sun totally eclipsed by the new moon. The new moon in silhouette forms the pupil, from the Latin "pupilla", for the "little doll" or "puppet" that you see when you look closely into the pupil of an eye, a tiny reflection of yourself. This is the *Naked Eye* of the Puppeteer, a sort of cosmic person who also represents nature, order, knowledge, and teacher. We are the pupils, the twinkle in the Puppeteer's eye.

Das Auge in der Hand ist die Sonne völlig durch den neuen Mond verfinstert. Der neue Mond in der Silhouette bildet die Schüler, aus dem lateinischen „pupilla", für die „kleine Puppe" oder „Marionette", die man sieht, wenn man genau in die Pupille eines Auges aussehen, eine winzige Reflexion von sich selbst. Dies ist für das bloße Auge des Puppeteer, eine Art kosmischer Person, die auch die Natur, Ordnung, Wissen und Lehrer darstellt. Wir sind die Schüler, die Zwinkern im Auge des Puppeteer.

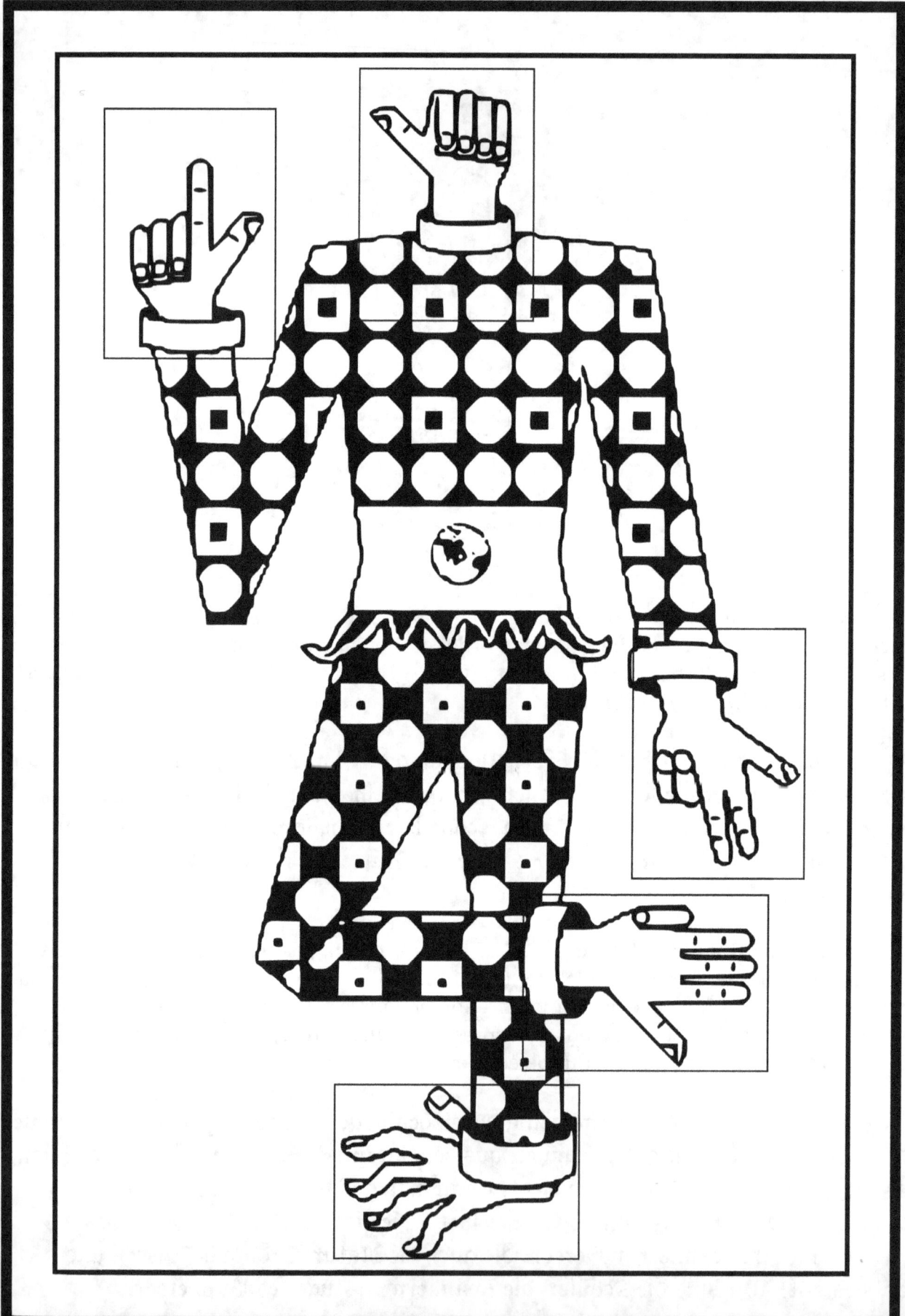

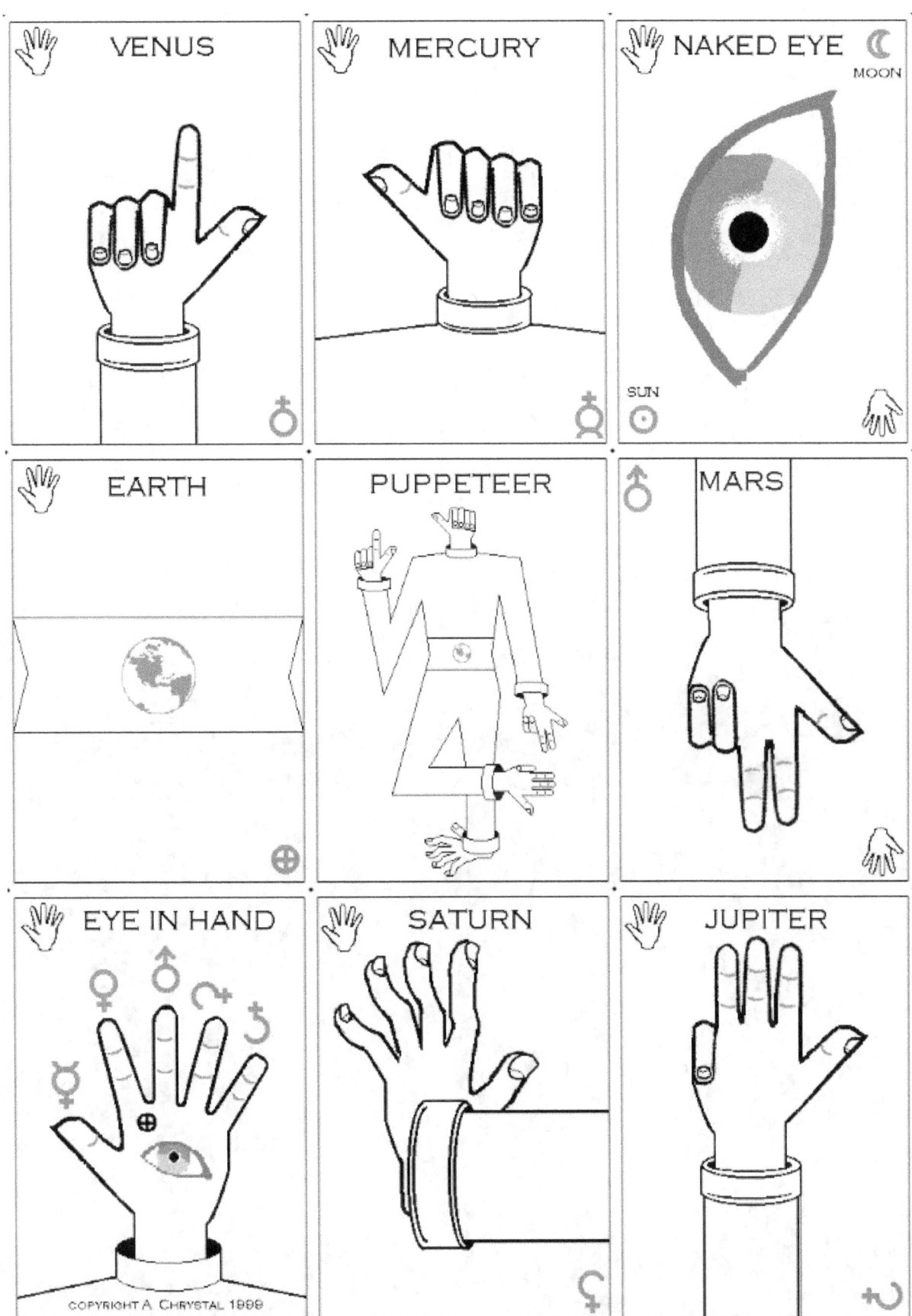

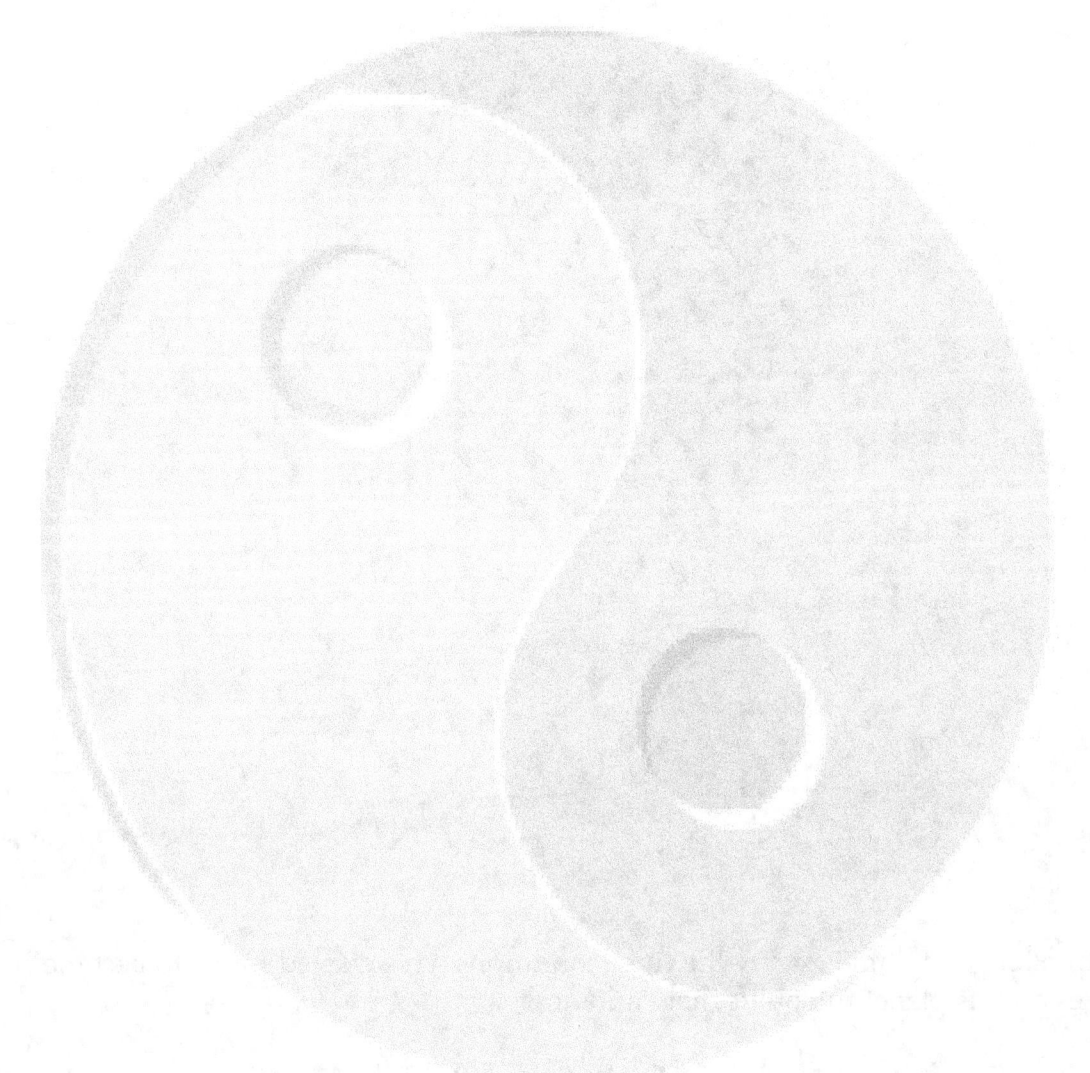

Warum?

Evolution

Because for a long long time exactly the right conditions have existed for us to become conscious of ourselves on this planet full of life, here and now.

Denn für eine lange lange Zeit genau die richtigen Bedingungen bestanden haben für uns auf diesem Planeten voller Leben bewusst selbst zu werden, hier und jetzt.

Incorporating the influence of the sun and moon, life as we know it has evolved on land and sea over vast stretches of time.

Die Einbeziehung des Einflusses der Sonne und des Mondes, das Leben wie wir es kennen auf Land und Meer über weite Strecken der Zeit entwickelt hat.

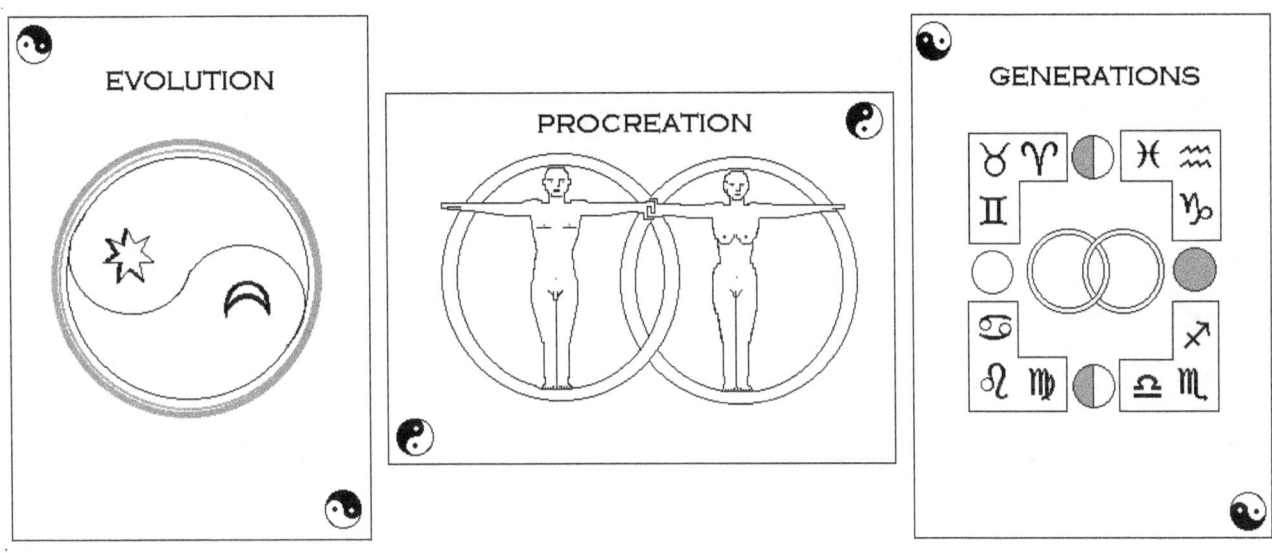

The interlocking rings represent the union of male and female that perpetuates human life by natural reproduction and parenting.

Die ineinander verschlungene Ringe repräsentieren die Vereinigung der männlichen und weiblichen dass verewigt das menschliche Leben durch natürliche Fortpflanzung und Elternschaft.

Time-honored convention associates the life span with the round of seasons and the lunar cycle. To illustrate this association *Generations* positions the moon's quarter phases roughly where the sun appears on the ecliptic at the solstices and equinoxes. Solstices are when the days are shortest or longest. Equinoxes are when days and nights are equal in length.

Altehrwürdige Konvention ordnet die Lebensdauer mit der Runde der Jahreszeit und dem Mondzyklus. Um dies zu verdeutlichen Assoziation Generationen Positionen der Viertelmondphasen etwa dort, wo die Sonne an den Sonnenwenden und Tagundnachtgleichen auf die Ekliptik erscheint. Solstices sind, wenn die Tage kürzesten oder längsten sind. Tagundnachtgleichen sind, wenn die Tage und Nächte gleich lang sind.

Wer?

Generationen

Life goes on generation after generation. A long life is likely to include 4 "hoods": childhood, parenthood, grandparenthood, and greatgrandparenthood. These are associated with 4 weeks in a month and 4 seasons in a year. The beginning and end of a lifetime are associated with the new moon, which is usually hidden in darkness and only revealed during a solar eclipse.

Das Leben geht weiter von Generation zu Generation. Ein langes Leben ist wahrscheinlich 4 „Hauben" enthalten: Kindheit, Elternschaft, Großelternschaft und greatgrandparenthood. Diese werden im Zusammenhang mit 4 Wochen in einem Monat und 4 Jahreszeit in einem Jahr. Der Anfang und das Ende des Lebens sind mit dem neuen Mond verbunden, die in der Regel in der Dunkelheit verborgen ist und nur während einer Sonnenfinsternis enthüllt.

We all share earth's biosphere, month after month, year after year, as long as we live.

Wir alle Biosphäre Anteil Erde, Monat für Monat, Jahr für Jahr, so lange wir leben.

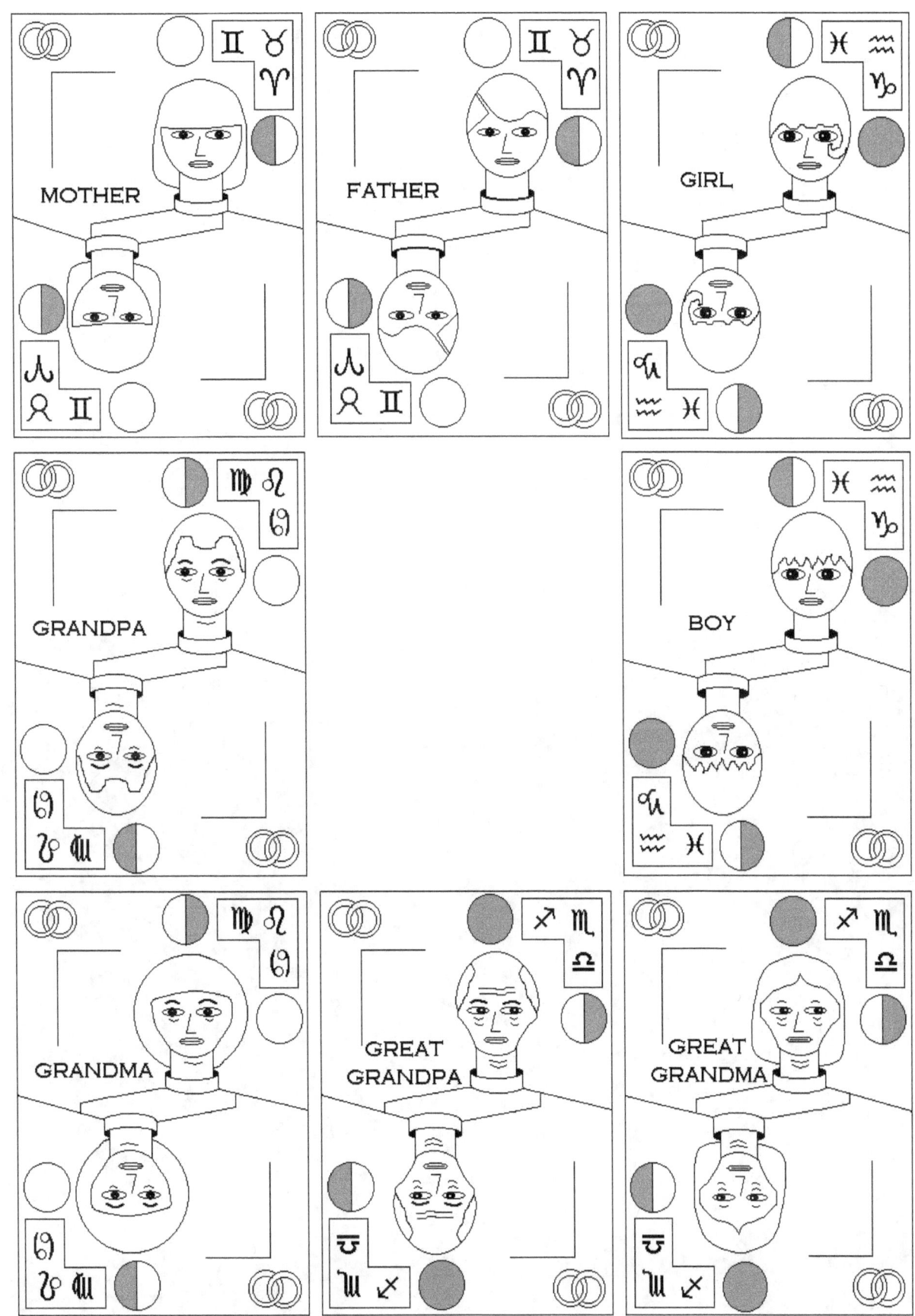

Was?

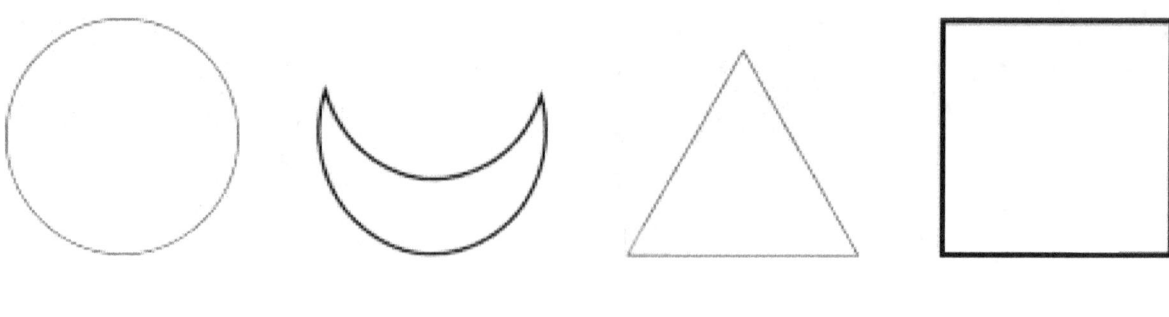

Elemente

The old theory that all things are made out of the four elements *Earth, Water, Fire*, and *Air* is true enough as far as it goes. Some ancients organized their thinking this way while others proposed an atomic theory.

Die alte Theorie, dass alle Dinge aus den vier Elementen aus Erde, Wasser, Feuer und Luft ist wahr genug, so weit wie es geht. Einige Alte organisierten ihr Denken auf diese Weise, während andere eine atomare Theorie vorgeschlagen.

The symbols for the elements used here are from eastern traditions. They are constructed of 1, 2, 3 or 4 circular or straight lines. This order is the basis for associating the elements and other components of the Puzzle such as the 4 seasons, and the 4 generations, as well as the 4 suits in a standard deck of 52 playing cards.

Die Symbole für die Elemente hier sind aus dem östlichen Tradition verwendet. Sie sind aus 1, 2 aufgebaut, 3 oder 4 kreisförmig oder gerade Linien. Die Bestellung ist die Grundlage für die Elemente und andere Komponenten des Puzzle wie die 4 Jahreszeit assoziieren, und die 4 Generationen sowie die 4 Farben in einem Standard-satz aus 52 Karten.

Note that the Puzzle uses the crescent both as a symbol for the element *Air*, and as a sign for the moon.

Man beachte, dass das Puzzle das Crescent verwendet sowohl als Symbol für das Element Luft, und als ein Zeichen für den Mond.

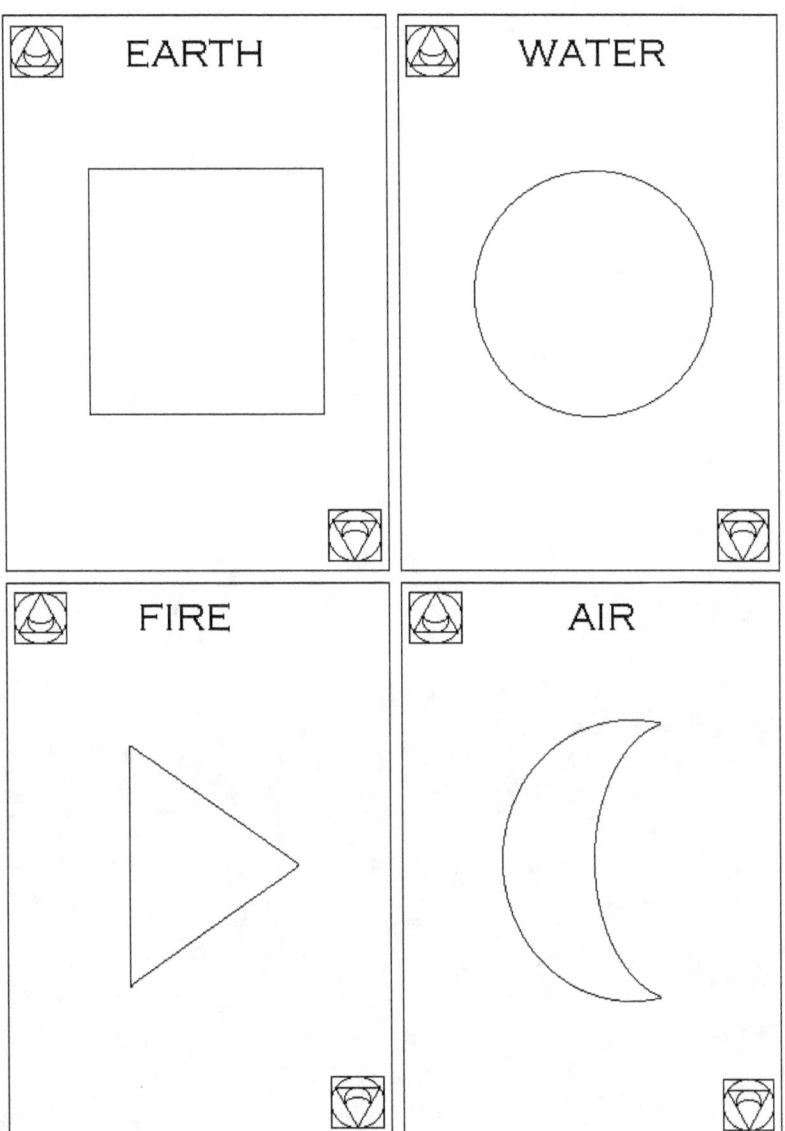

Wann?

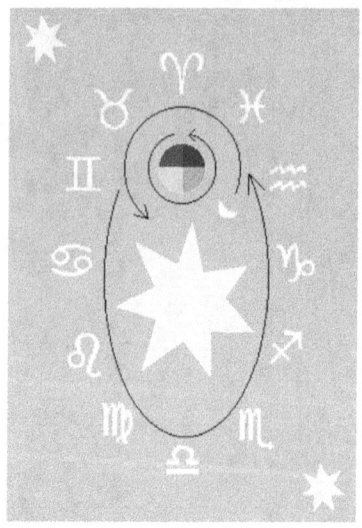

Kalender

Everything takes time. When objects move through space at a constant rate the distance traveled is analogous to the time elapsed, like the hands of a mechanical clock. The hour hand of a 24-hour clock mimics the sun's apparent daily motion through the sky.

Alles braucht seine Zeit. Wenn Objekte durch den Raum mit einer konstanten Geschwindigkeit bewegen, die zurückgelegte Strecke ist analog zu der verstrichenen Zeit, wie die Hände einer mechanischen Uhr. Der Stundenzeiger einer 24-Stunden-Uhr ahmt die scheinbare tägliche Bewegung durch den Himmel der Sonne.

A few centuries ago it became widely accepted that the apparent motions of the sun and moon are due to the daily rotation of the earth on its axis, the monthly orbit of the moon around the earth, and the yearly orbit of the earth/moon pair around the sun, all turning and spinning in the same direction.

Vor einigen Jahrhunderten wurde es allgemein anerkannt, dass die scheinbaren Bewegungen der Sonne und des Mondes auf die tägliche Rotation der Erde um ihre Achse zurückzuführen sind, die monatliche Umlaufbahn des Mondes um die Erde, und der jährlichen Umlaufbahn der Erde / Mond Paar um die Sonne, die alle Dreh- und Spinnen in der gleichen Richtung.

The moon goes through its phases in about 28 days. A quarter of the lunar cycle is about 7 days, one week. The days of the week are associated with the sun, moon, and five planets visible to the naked eye. To order them fastest to slowest by angular velocity (their apparent speed relative to the fixed stars), start with the moon and go counter clockwise skipping every other one: Moon, Mercury, Venus, Sun, Mars, Jupiter, Saturn.

Der Mond geht durch Phasen in etwa 28 Tagen. Ein Viertel des Mondzyklus ist ungefähr 7 Tage, eine Woche. Die Tage der Woche mit der Sonne, den Mond verbunden ist, und fünf Planeten mit bloßem Auge. Um sie am schnellsten zum langsamsten durch Winkelgeschwindigkeit (ihre scheinbare Geschwindigkeit relativ zu den Fixsternen) zu bestellen, beginnen mit dem Mond und gehen gegen den Uhrzeigersinn alle anderen ein Überspringen: Mond, Merkur, Venus, Sonne, Mars, Jupiter, Saturn.

Lunar and solar cycles are not whole numbers of days in duration. A solar cycle is not a whole number of lunar cycles. The most accurate calendars have 365 days in most years. Another day is still needed every so often to catch the calendar year up with the actual time it takes for the sun to return to the same point on the ecliptic.

Mond- und Sonnenzyklen sind nicht ganze Anzahl von Tagen Dauer. Ein Sonnenzyklus ist nicht eine ganze Anzahl von Mondzyklen. Die genauesten Kalender haben 365 Tage in den meisten Jahren. Ein weiterer Tag ist noch jeder so oft erforderlich, um das Kalenderjahr, um aufzuholen mit der tatsächlichen Zeit es braucht, um die Sonne auf den gleichen Punkt auf der Ekliptik zurückzukehren.

When the apparent sizes of the sun and moon are identical and they intersect perfectly on the ecliptic, the sun's dazzling white corona appears to encircle the new moon. You may see planets and bright stars during the day, and wonderful colors circling the horizon. On earth it can be viewed with the naked eye during totality which may last just a few moments, or more than 7 minutes.

Wenn die scheinbare Größe der Sonne und des Mondes identisch sind und sie sich schneiden perfekt auf der Ekliptik, erscheint der Sonne blendend weiße Korona den Neumond zu umgeben. Sie können Planeten und helle Sterne im Laufe des Tages sehen, und wundervolle Farben den Horizont kreisen. Auf der Erde kann es mit dem bloßen Auge während der Totalität betrachtet werden, die nur ein paar Minuten dauern können, oder mehr als 7 Minuten.

The timing and positioning of a total *Solar Eclipse* depends on the combined motions of the moon, the sun, and the earth. Predicting them has been a preoccupation of astronomers for millennia. Total solar eclipses don't repeat neatly in any one earthly location, but often recur after about 54 years, and about 1000 kilometers west of one in the same Saros series. Ancient records of eclipses have been used to precisely date events in the remote past when other means are inconclusive.

Das Timing und die Positionierung einer Sonnenfinsternis hängt von den kombinierten Bewegungen des Mondes, der Sonne und der Erde. Die Vorhersage sie hat eine Beschäftigung von Astronomen seit Jahrtausenden. Insgesamt Sonnenfinsternisse nicht wiederholt säuberlich in einer irdischen Lage, aber oft nach etwa 54 Jahren wiederkehren, und etwa 1000 Kilometer westlich von einem in der gleichen Saros-Serie. Alte Aufzeichnungen von Finsternissen wurden genau Datum Ereignisse in der fernen Vergangenheit verwendet, wenn andere Mittel nicht schlüssig sind.

1ST QUARTER

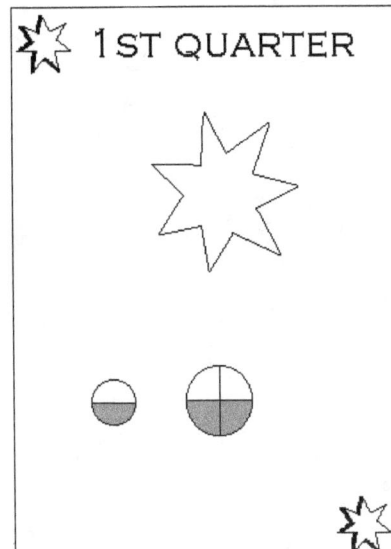

YEAR

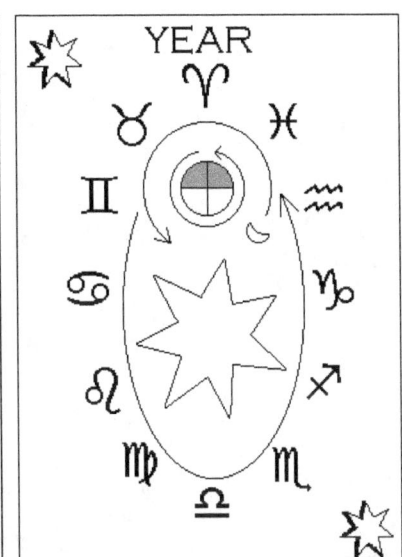

NEW MOON

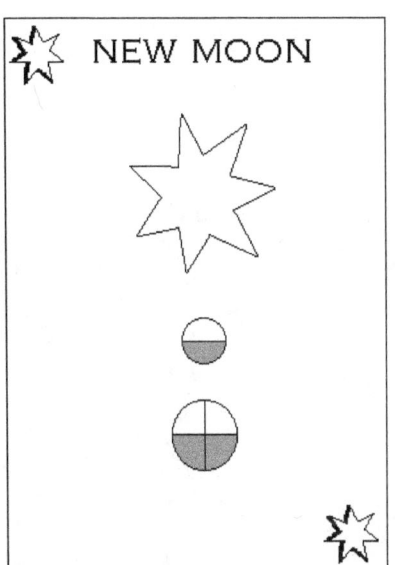

MONTH

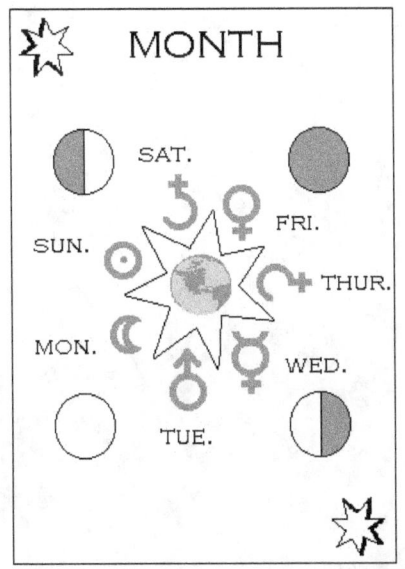

FULL MOON

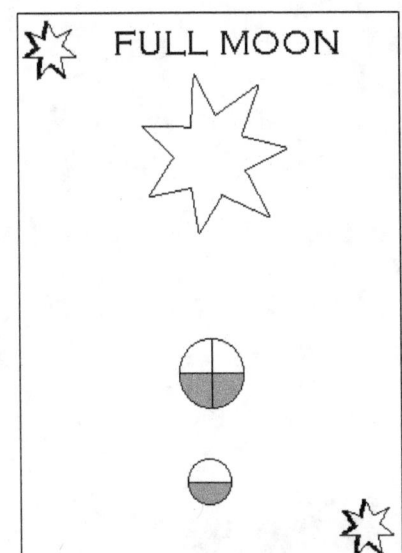

SOLAR ECLIPSE

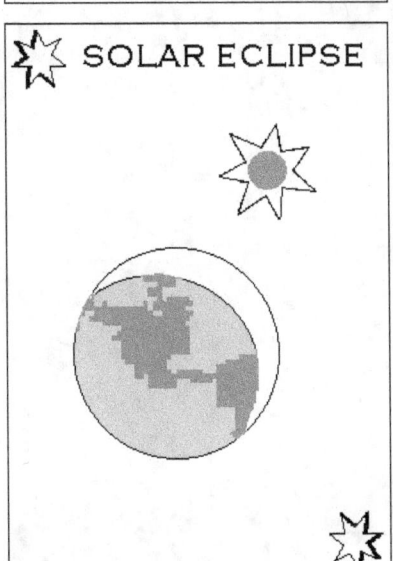

3RD QUARTER

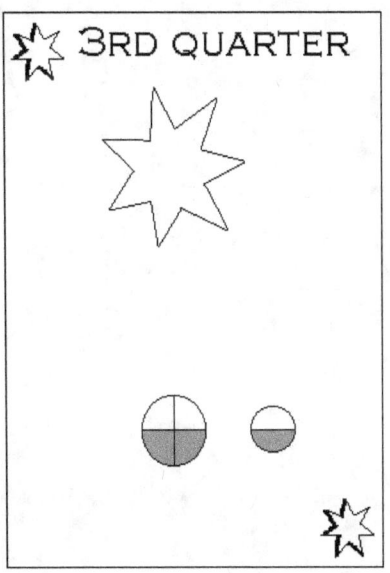

Wo?

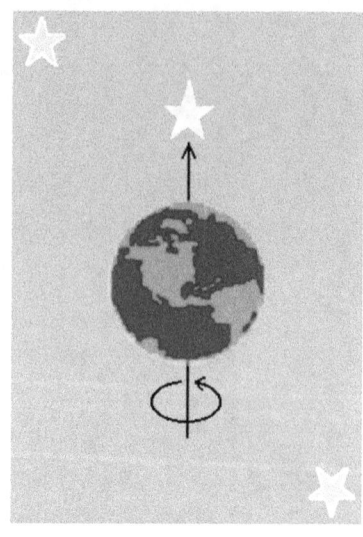

Kompass

At night the sky full of heavenly bodies appears to slowly circle the poles. In the north there is one pole star that doesn't appear to move. The sun and often the moon appear to circle the same way during the day, rising in the east and setting in the west. This apparent celestial circling motion is due to the rotation of the earth around its *Polar Axis*, west to east.

In der Nacht erscheint der Himmel voller Himmelskörper langsam den Pol kreisen zu. Im Norden gibt es einen Pol Stern, der sich nicht zu bewegen scheint. Die Sonne und erscheinen oft der Mond zu umkreisen die gleiche Art und Weise während des Tages, im Osten aufgehen und im Westen unter. Diese scheinbare Himmelskreisbewegung ist aufgrund der Drehung der Erde um ihre Polarachse, von Westen nach Osten.

A crescent moon associated with a setting sun is a waxing moon. A crescent moon associated with a rising sun is a waning moon.

Ein Halbmond mit einer untergehenden Sonne verbunden ist, ist ein zunehmender Mond. Ein Halbmond mit einer steigenden Sonne verbunden ist abnehmender Mond.

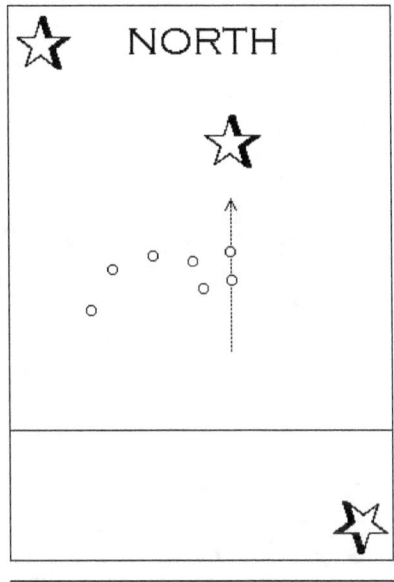

NORTH

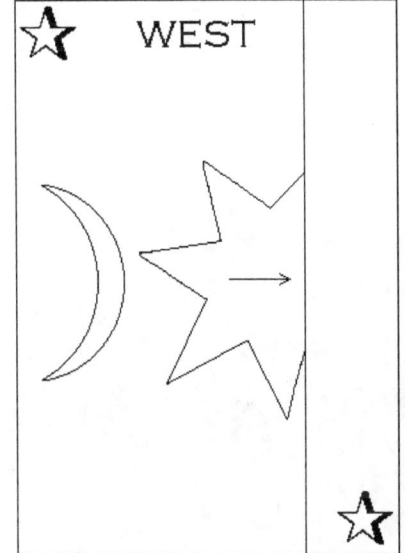

WEST

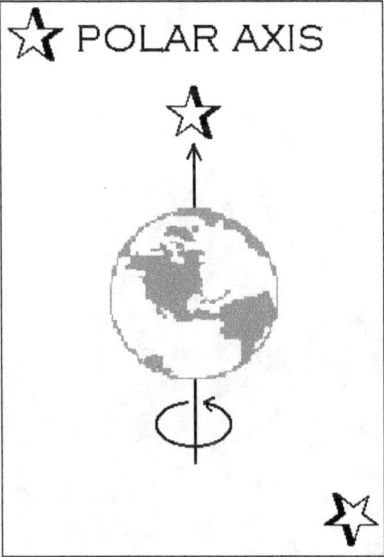

POLAR AXIS

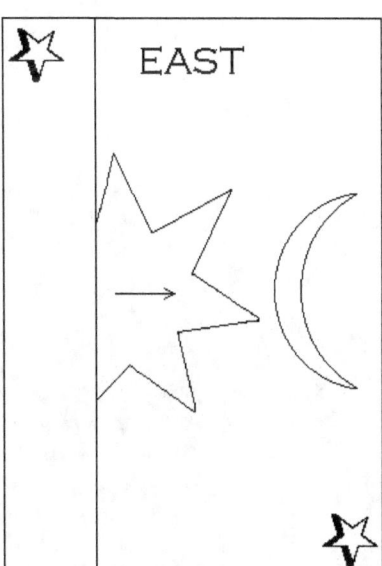

EAST

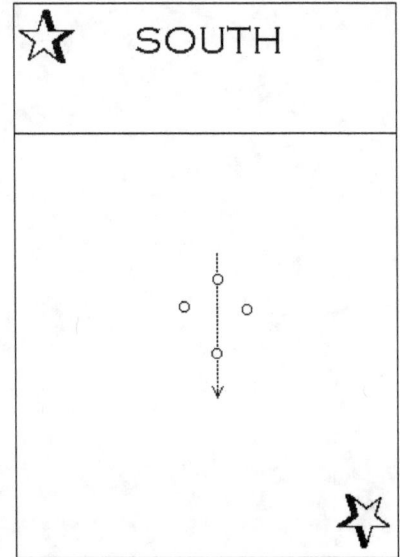

SOUTH

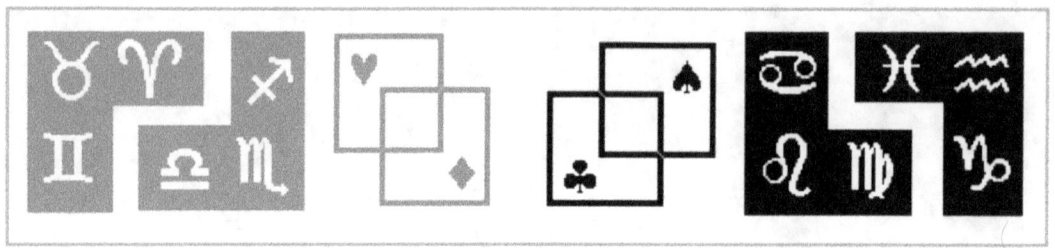

Die Ekliptik

Every year the sun appears to retrace a path through the stars, called the ecliptic. The apparent motions of the moon and the planets are confined to a region near the ecliptic. Solar and lunar eclipses only happen when the moon crosses the ecliptic.

Jedes Jahr erscheint die Sonne einen Weg durch die Sterne zurückzuverfolgen, die Ekliptik genannt. Die scheinbaren Bewegungen des Mondes und der Planeten zu einem Bereich nahe der Ekliptik beschränkt. Sonnen- und Mondfinsternisse nur geschehen, wenn der Mond die Ekliptik kreuzt.

The moon goes through its phases about 12 times in a year prompting the division of the sun's progress along the ecliptic into twelve sections. These divisions are roughly marked by the constellations of the zodiac. They were devised over 5,000 years ago.

Der Mond geht seine Phasen etwa 12 Mal im Jahr auffordert, die Teilung der Sonnen fortschritts entlang der Ekliptik in zwölf Abschnitte durch. Diese Bereiche sind in etwa von den Konstellationen des Tierkreises markiert. Sie wurden vor 5000 Jahren erdacht über.

The 52 cards of a standard deck are divided into 4 equal suits, like a year of 52 weeks divided into 4 seasons. One common convention ranks the suits high to low: spades, hearts, clubs, diamonds, the order used on the Puzzle card back shown on the back cover. Also shown on the back cover, the 12 Ecliptic cards comprise two suits: one associated with the black suits of a standard deck and one associated with the red suits.

Die 52 Karten eines Standard-satz sind in 4 gleiche Farben unterteilt, wie ein Jahr 52 Wochen in vier Jahreszeiten unterteilt. Eine übliche Konvention stuft die Farben hoch zu niedrig: Pik, Herz, Kreuz, Karo, die Reihenfolge auf der Puzzle-Karte verwendet zurück auf der Rückseite dargestellt. Auch auf der Rückseite gezeigt, die 12 Ekliptik Karten umfassen zwei Farben: eine im Zusammenhang mit den schwarzen Farben von einem Standard-satz und einem mit den roten Farbenen verbunden.

The sun appears in these constellations on these dates:

Die Sonne scheint in diesen Konstellationen an diesen Tagen:

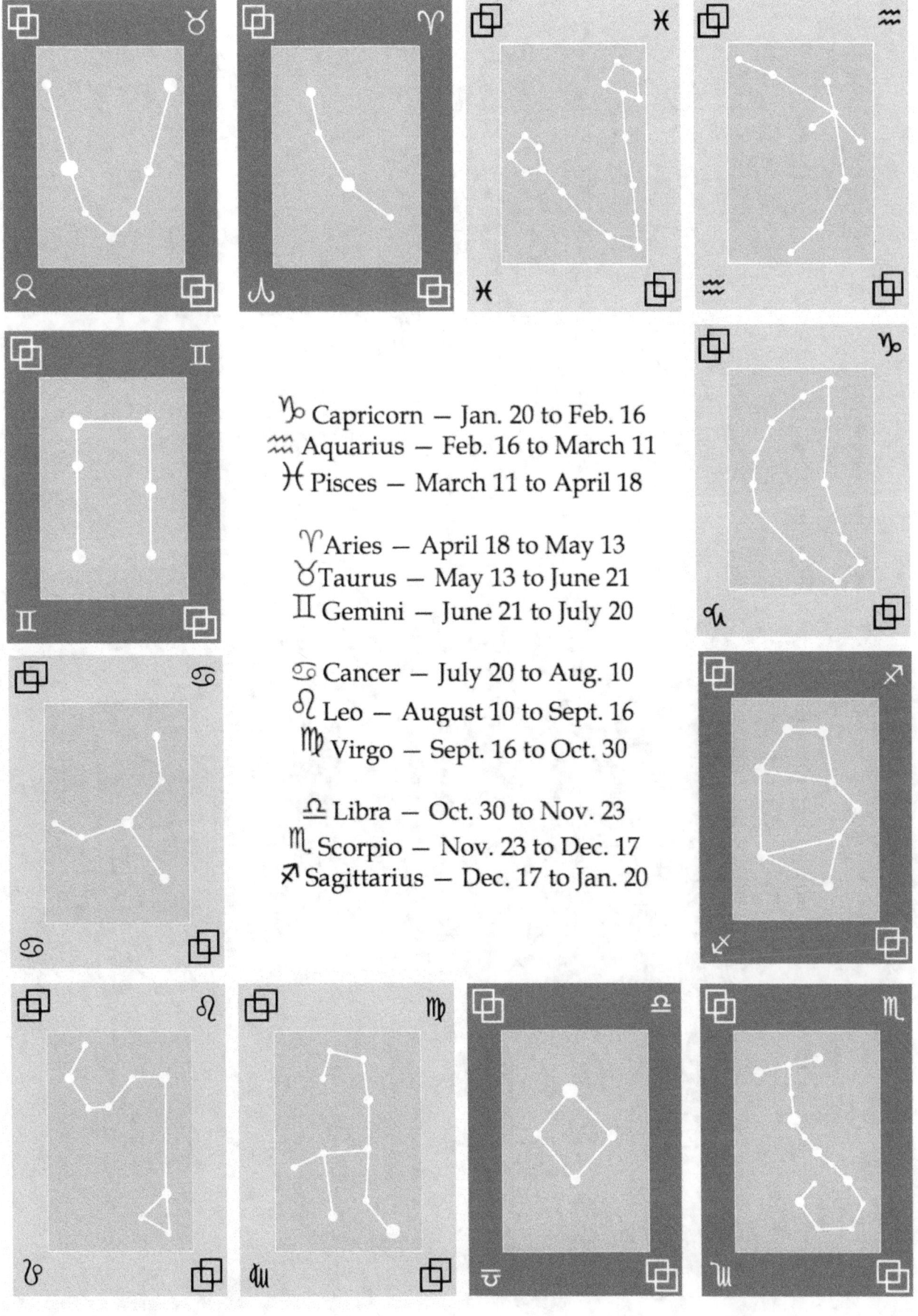

♑ Capricorn — Jan. 20 to Feb. 16
♒ Aquarius — Feb. 16 to March 11
♓ Pisces — March 11 to April 18

♈ Aries — April 18 to May 13
♉ Taurus — May 13 to June 21
♊ Gemini — June 21 to July 20

♋ Cancer — July 20 to Aug. 10
♌ Leo — August 10 to Sept. 16
♍ Virgo — Sept. 16 to Oct. 30

♎ Libra — Oct. 30 to Nov. 23
♏ Scorpio — Nov. 23 to Dec. 17
♐ Sagittarius — Dec. 17 to Jan. 20

♐ ♏ ♍ ♎ ♍ ♌ ♋ ♊ ♉ ♈ ♓ ♒

♑ ♐ ♏ ♍ ♎ ♍ ♌ ♋ ♊ ♉ ♈ ♓

♒ ♑ ♐ ♏ ♍ ♎ ♍ ♌ ♋ ♊ ♉ ♈

♓ ♒ ♑ ♐ ♏ ♍ ♎ ♍ ♌ ♋ ♊ ♉

♈ ♓ ♒ ♑ ♐ ♏ ♍ ♎ ♍ ♌ ♋ ♊

♉ ♈ ♓ ♒ ♑ ♐ ♏ ♍ ♎ ♍ ♌ ♋

♊ ♉ ♈ ♓ ♒ ♑ ♐ ♏ ♍ ♎ ♍ ♌

♋ ♊ ♉ ♈ ♓ ♒ ♑ ♐ ♏ ♍ ♎ ♍

♌ ♋ ♊ ♉ ♈ ♓ ♒ ♑ ♐ ♏ ♍ ♎

♍ ♌ ♋ ♊ ♉ ♈ ♓ ♒ ♑ ♐ ♏ ♍

♎ ♍ ♌ ♋ ♊ ♉ ♈ ♓ ♒ ♑ ♐ ♏

♏ ♍ ♎ ♍ ♌ ♋ ♊ ♉ ♈ ♓ ♒ ♑

♐ ♏ ♍ ♎ ♍ ♌ ♋ ♊ ♉ ♈ ♓ ♒

♑ ♐ ♏ ♍ ♎ ♍ ♌ ♋ ♊ ♉ ♈ ♓

Spiele

As play things and conversation pieces, Cosmic Puzzle cards can be introduced around age two. Older playmates and caregivers are encouraged to use them with preverbal children for free play or spontaneously structured games. Many familiar games played with a standard deck can be adapted for the Cosmic Puzzle.

Als Spiel Dinge und Unterhaltung Stücke können kosmische Puzzle Karten etwa im Alter von zwei eingeführt werden. Ältere Gespielinnen und Betreuer werden ermutigt, sie mit preverbal Kinder für freies Spiel oder spontan strukturierte Spiele zu verwenden. Viele bekannten Spiele mit einem Standard-Satz gespielt für das kosmische Puzzle angepasst werden.

Becoming familiar with the cards is the first step towards using them for conventional card play. The card back is pictured on the back cover and shows how the 52 spades, hearts, clubs, and diamonds, the **12 Ecliptic cards**, and the other **36 Puzzle cards** all fit neatly into a 10 x 10 grid. The 36 are represented by the color of their 6 suit symbols; this provides a quick reference for how many cards are in each suit. The fronts of the 36 are on the front cover. The fronts of the 12 are on the back cover.

Wird vertraut mit den Karten ist der erste Schritt auf dem Weg sich für herkömmliches Kartenspiel verwenden. Die Kartenrückseite ist auf der Rückseite dargestellt und zeigt, wie die 52 Pik, Herz, Kreuz, Karo und die 12 Ekliptik-Karten, und die anderen 36 Puzzle-Karten alle paßt genau in ein 10 x 10 Raster. Die 36 sind durch die Farbe ihrer 6-Farben Symbolen dargestellt; dies stellt eine schnelle Referenz für wie viele Karten in jeder Farbe sind. Die Fronten der 36 sind auf der vorderen Abdeckung. Die Fronten der 12 sind auf der Rückseite.

Included in the Millennium Edition, but not shown on the book cover are **6 Text cards**, which give rhyming clues in English about the meaning of each question suit. The 6 Text cards may be included in their respective suits, used as spares, or wild cards. A **Title card** is included, too, but not shown on the book cover. It is best used as a wild card or set aside for use as a spare.

Eingeschlossen in der Millennium Edition, aber nicht auf dem Buchdeckel sind 6 Text Karten gezeigt, die in englischer Sprache gereimten Hinweise geben über die Bedeutung der einzelnen Fragen Farbe. Die 6 Text-Karten können in ihren jeweiligen Klagen aufgenommen werden, als Ersatzteile verwendet werden, oder Wildcards. Eine Titelkarte enthalten ist, auch, aber nicht auf dem Buchdeckel gezeigt. Es ist am besten als Platzhalter oder beiseite legen für die Verwendung als Ersatz verwendet.

Matrix

The 12 Ecliptic cards, the other 36 Puzzle cards, and the 6 Text cards are arranged on six 3x3 grids, as shown below. The same grids are printed two to a page on following pages. Those can be cut out or copied and used in play. Grids are distributed equally, one or more to a player.

Die 12 Ekliptik-Karten, die anderen 36 Puzzle-Karten, und die 6 Textkarten werden auf sechs 3x3-Gitter angeordnet sind, wie unten gezeigt. Die gleichen Gitter sind zwei auf einer Seite auf den folgenden Seiten gedruckt. Diese können im Spiel ausgeschnitten oder kopiert und verwendet werden. Gitter sind gleichmäßig verteilt, eine oder mehrere an einen Spieler.

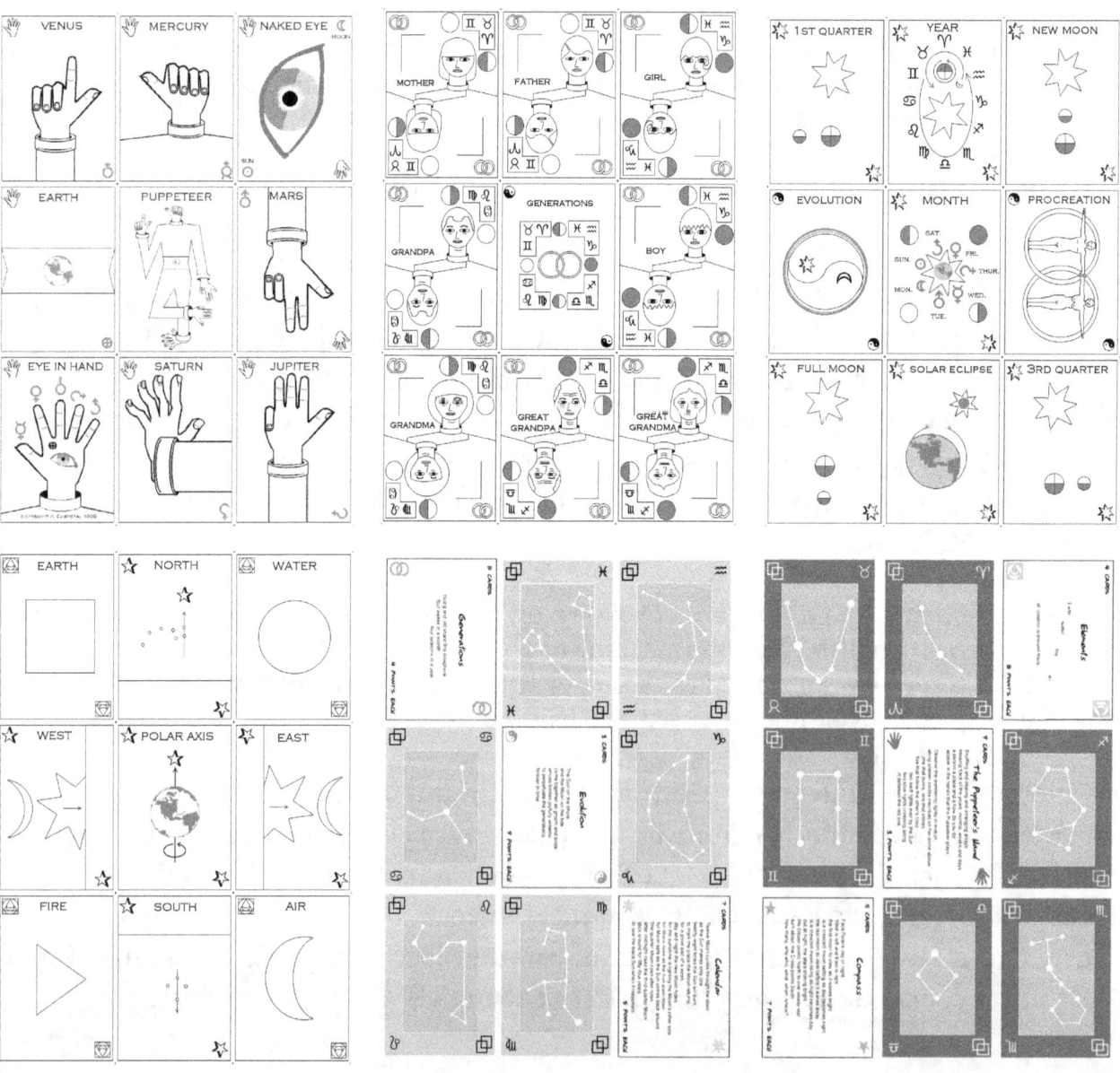

Each player starts with 9 tokens (such as pennies) for each grid being played. The deck is shuffled well and placed face down. Cards are turned up one at a time. If a card appears on a player's grid, a token is placed on the space where it appears. The first player to get 3 tokens in a line (row, column, or diagonal) and declare a "matrix" wins the round. All players surrender the tokens they have placed so far to the winner of the round. Between rounds players exchange 1 or more grids and the whole deck is shuffled well. Play continues until any player has fewer than 3 tokens between rounds, or some other agreed upon time limit is reached, or number of rounds has been played. The player with the most tokens wins.

Jeder Spieler beginnt mit 9 Tokens (wie ein paar Cent) für jedes Raster gespielt wird. Das Deck ist gut gemischt und legte das Gesicht nach unten. Die Karten werden einer nach dem anderen aufgedreht. Wenn eine Karte auf einem Raster des Spielers angezeigt wird, wird ein Token auf dem Raum platziert, wo es angezeigt wird. Der erste Spieler, zu erhalten 3 Token in einer Linie (Zeile, Spalte oder diagonal) und erklärt, eine „Matrix", um die Runde gewinnt. Alle Spieler übergeben die Token sie bisher an den Gewinner der Runde gesetzt haben. Zwischen den Runden Spieler tauschen 1 oder mehrere Gitter und das ganze Deck ist gut gemischt. Das Spiel wird fortgesetzt, bis ein Spieler weniger als 3 Token zwischen den Runden hat, oder eine andere auf Frist vereinbart erreicht ist, oder die Anzahl der Runden gespielt wurde. Der Spieler mit den meisten Token gewinnt.

3D Matrix

Similar to Matrix, but for 2 players each playing 3 grids, as if stacked in layers. Before each round players take turns choosing and arranging new grids. In addition to row, column, and diagonal alignments on a single layer, 3 tokens may be lined up one layer to the next as if stacked vertically or diagonally.

Ähnlich wie bei Matrix, aber für 2 Spieler jeden Spiel 3 Roste, als ob in Schichten gestapelt. Vor jeder Runde reihum Auswahl und neue Gitter angeordnet werden. Zusätzlich zu Reihe, Spalte und Diagonale Ausrichtungen auf einer einzelnen Schicht, 3 Token einer Schicht zur nächsten aufgereiht werden können, als ob gestapelt vertikal oder diagonal.

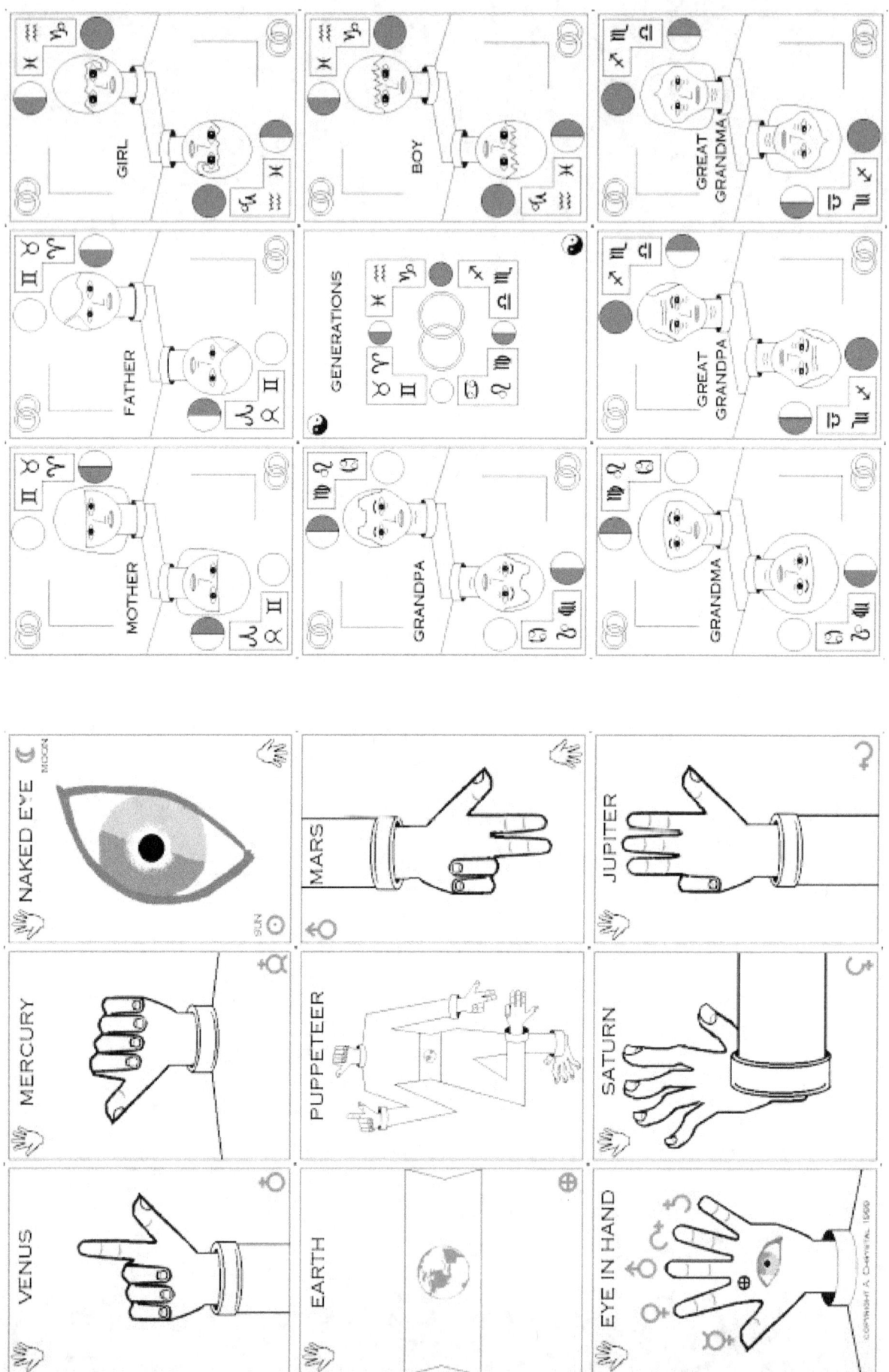

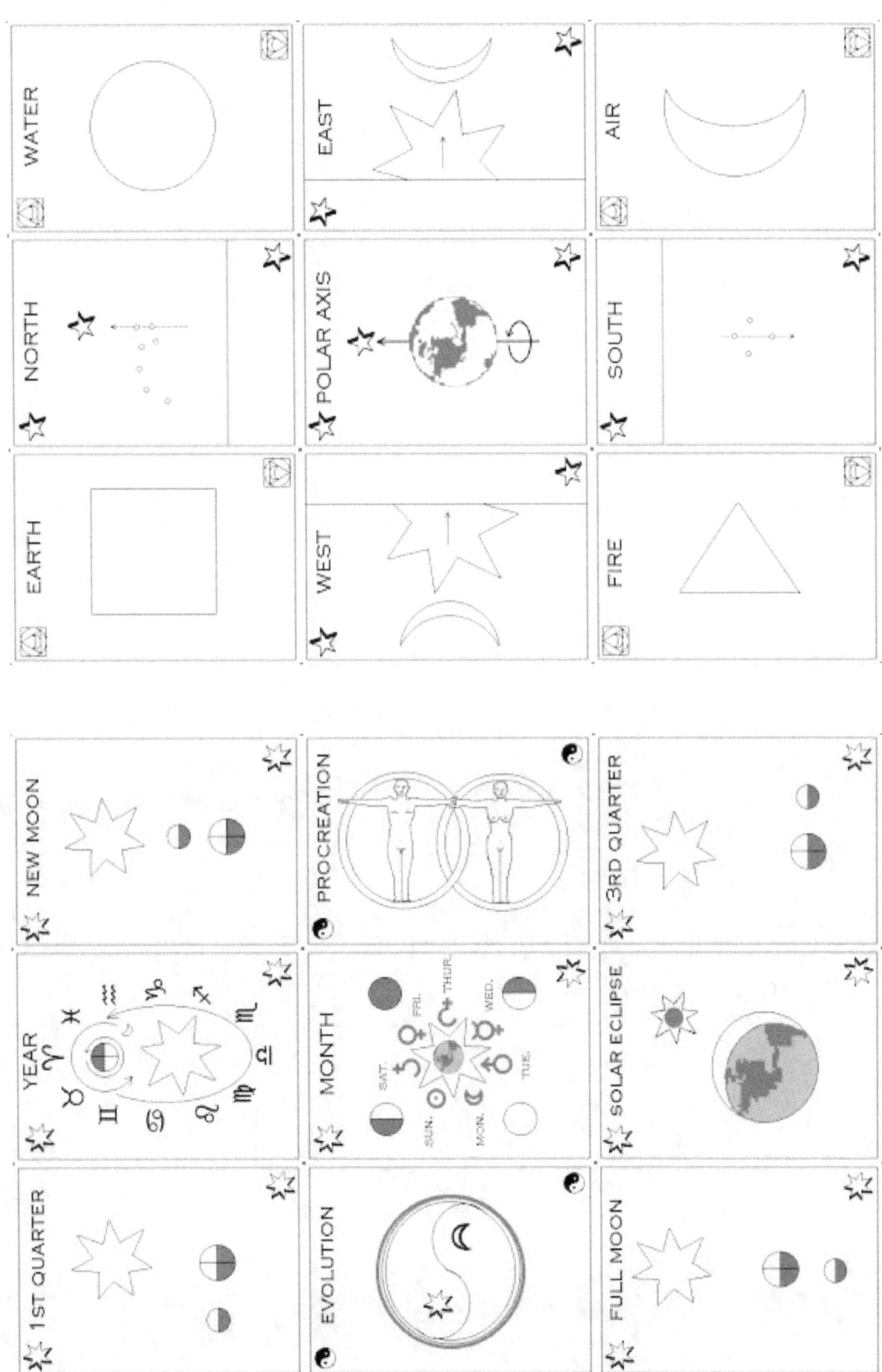

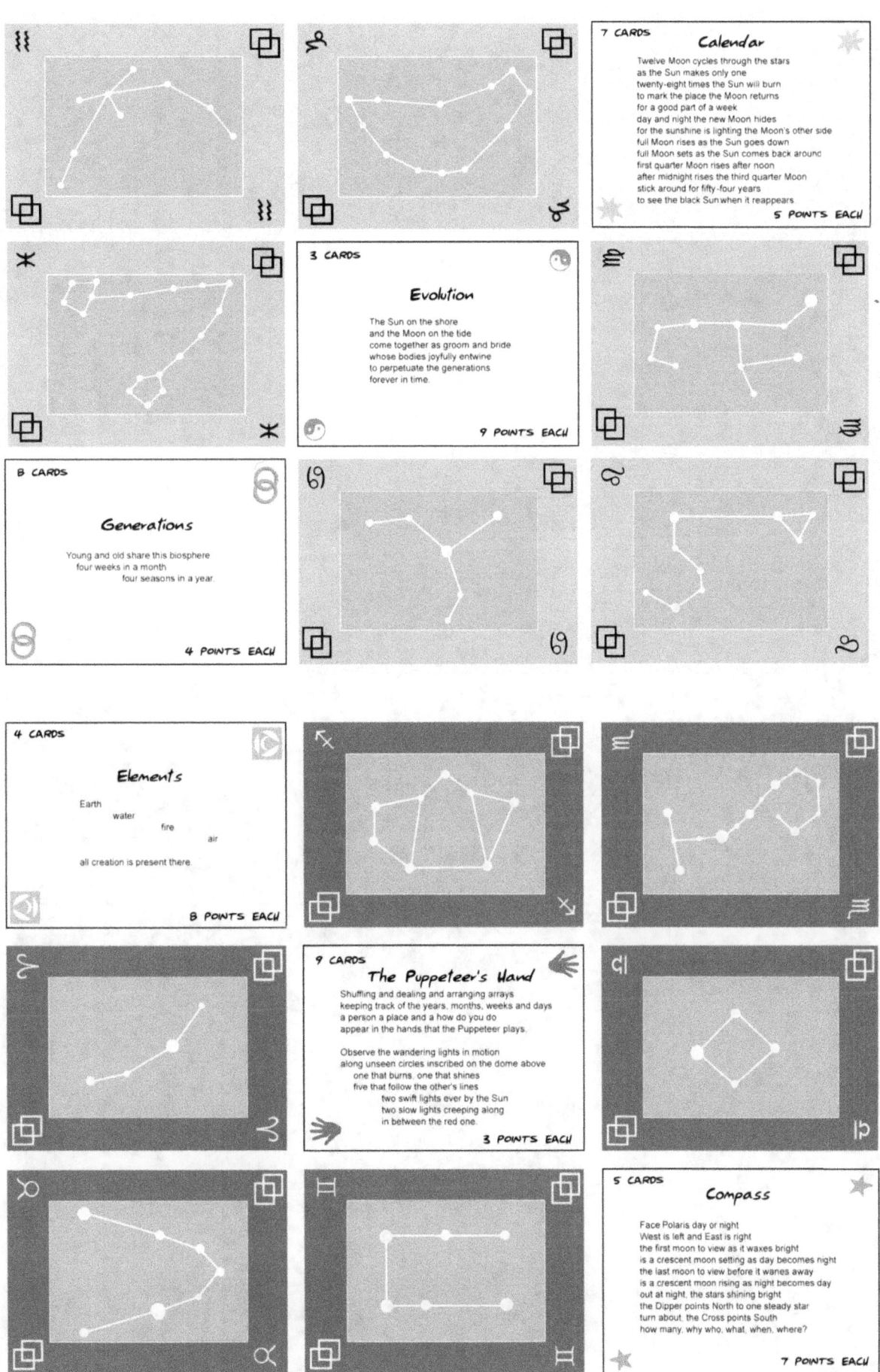

7 CARDS — Calendar

Twelve Moon cycles through the stars
as the Sun makes only one
twenty-eight times the Sun will burn
to mark the place the Moon returns
for a good part of a week
day and night the new Moon hides
for the sunshine is lighting the Moon's other side
full Moon rises as the Sun goes down
full Moon sets as the Sun comes back around
first quarter Moon rises after noon
after midnight rises the third quarter Moon
stick around for fifty-four years
to see the black Sun when it reappears

5 POINTS EACH

3 CARDS — Evolution

The Sun on the shore
and the Moon on the tide
come together as groom and bride
whose bodies joyfully entwine
to perpetuate the generations
forever in time.

9 POINTS EACH

8 CARDS — Generations

Young and old share this biosphere
four weeks in a month
four seasons in a year.

4 POINTS EACH

4 CARDS — Elements

Earth
water
fire
air

all creation is present there.

8 POINTS EACH

9 CARDS — The Puppeteer's Hand

Shuffling and dealing and arranging arrays
keeping track of the years, months, weeks and days
a person a place and a how do you do
appear in the hands that the Puppeteer plays.

Observe the wandering lights in motion
along unseen circles inscribed on the dome above
one that burns, one that shines
five that follow the other's lines
two swift lights ever by the Sun
two slow lights creeping along
in between the red one.

3 POINTS EACH

5 CARDS — Compass

Face Polaris day or night
West is left and East is right
the first moon to view as it waxes bright
is a crescent moon setting as day becomes night
the last moon to view before it wanes away
is a crescent moon rising as night becomes day
out at night, the stars shining bright
the Dipper points North to one steady star
turn about, the Cross points South
how many, why who, what, when, where?

7 POINTS EACH

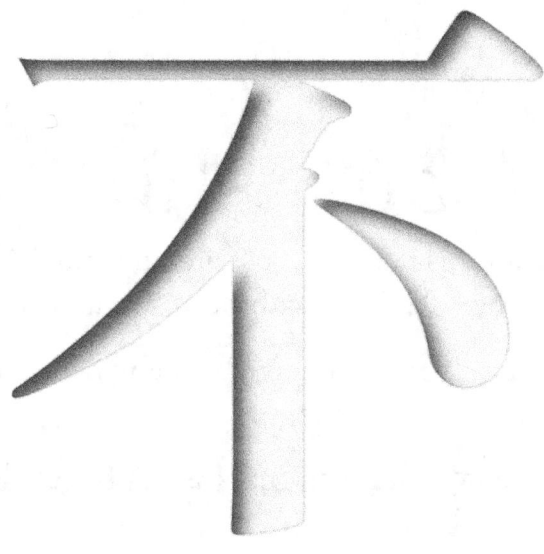

bú shì

Meaning "is not", or "no"; literally "not is".

𝔅edeutung „ist nicht“ oder „𝔑ein“; wahrsten 𝔖inne des 𝔚ortes „nicht ist“.

This sort of bluffing and challenging game is known by many names when played with standard playing cards.

𝔇iese 𝔄rt von bluffen und herausfordernden 𝔖piel ist unter vielen 𝔑amen bekannt, als mit 𝔖tandard-𝔖pielkarten gespielt.

For 3 or more players, the object is to be the first to get rid of all the cards in your hand. Remove the Title card and the 12 Ecliptic cards from the deck. Shuffle and deal the remaining 42 cards evenly to all players, using any cards left over to start the discard pile face down in the middle of the play area.

𝔉ür drei oder mehr 𝔖pieler, ist das 𝔒bjekt, die erste, um loszuwerden, alle 𝔎arten in der ℌand zu sein. 𝔈ntfernen 𝔖ie die 𝔗itelkarte und die 12 𝔈kliptik 𝔎arten aus dem 𝔇eck. 𝔖huffle und sich die restlichen 42 𝔎arten gleichmäßig an alle 𝔖pieler, indem die 𝔎arten links über den 𝔄blagestapel bedruckten 𝔖eite nach unten in der 𝔐itte des 𝔖pielfeldes zu starten.

Players take turns discarding onto the pile while verbally asserting the number and suit of the cards shed. Every player in turn must assert that they are discarding at least 1 card from the featured suit. The featured suits proceed through the rainbow. The 1st player asserts that they are discarding red🖐, the 2nd player asserts that they are discarding orange◯◯, the next yellow✦, then green☆, blue🔺, and violet☯. Then start again with red.

𝔇ie 𝔖pieler ziehen abwechselnd auf den 𝔖tapel zu verwerfen, während verbal die 𝔄nzahl und 𝔉arbe der 𝔎arten 𝔖chuppen zu behaupten. 𝔍eder 𝔖pieler muss wiederum behaupten, dass sie mindestens 1 𝔎arte aus dem vorgestellten 𝔉arbe sind zu verwerfen. 𝔇ie vorgestellten 𝔎lagen gehen durch die 𝔎egenbogen. 𝔇er erste 𝔖pieler behauptet, dass sie rot🖐 sind 𝔙erwerfen, behauptet der zweite 𝔖pieler, dass sie 𝔒range◯◯ sind zu verwerfen, die nächste gelb✦, dann grün☆, blau🔺 und violett☯. 𝔇ann beginnt wieder mit rot.

If one player suspects that any of another player's cards **IS NOT** what they say it is, they may challenge the other player by immediately declaring **"bú shì!"** before anyone else discards. The challenged player then has to turn face up all the cards they just discarded. If they were bluffing they have to add the entire discard pile to their hand. If they weren't bluffing the challenger must add the entire discard pile to their hand.

Wenn ein Spieler den Verdacht hegt, dass eines anderen Spielers Karten ist nicht, was sie sagen, es ist, können sie die anderen Spieler herausfordern, indem sie unverzüglich zu erklären „bu shi!", Bevor jemand anderes verwirft. Der angegriffene Spieler hat dann Gesicht all Karten drehen sie einfach verworfen. Wenn sie bluffen müssen sie den gesamten Ablagestapel auf ihre Hand hinzufügen. Wenn sie nicht den Herausforderer bluffen muss den gesamten Ablagestapel auf ihre Hand hinzufügen.

Solitar

Suitable for coloring and for solitary play such as sorting or matching activities, the 54 Puzzle and Text cards in the Millennium Edition of The Puppeteer's Cosmic Puzzle are presented below as grids of labeled line drawings, 9 cards to a page. These are nearly to scale and serve as a reference for learning the names of the cards as well as the meaning of the sun/moon/planet ciphers, and their associated weekdays.

Geeignet zum Färben und für einsames Spiel wie Sortieren oder Aktivitäten übereinstimmt, die 54 Puzzle und Text Karten in der Millennium Edition des kosmischen Puzzle des Puppeteer werden als Gitter markierter Linienzeichnungen unten dargestellt, 9 Karten auf einer Seite. Dies sind fast maßstab und dienen als Referenz für das Lernen, die Namen der Karten sowie die Bedeutung der Sonne / Mond / Planeten Chiffren und ihre zugehörigen Wochentag.

For other games, go to:
für andere Spiele, gehen Sie zu:

cosmicpuzzle.com/games.htm

For tips on pseudopsychic readings, go to:
Tipps zur pseudopsychic Lesungen, gehen Sie zu:

cosmicpuzzle.com/divination.htm

For the related board game Total Eclipse, go to:
für das damit verbundene Brettspiel Total Eclipse, gehen Sie zu:

thegamecrafter.com/games/total-eclipse

不是不是不是不

是不是不是不是

不是不是不是不

是不是不是不是

不是不是不是不

是不是不是不是

不是不是不是不

是不是不是不是

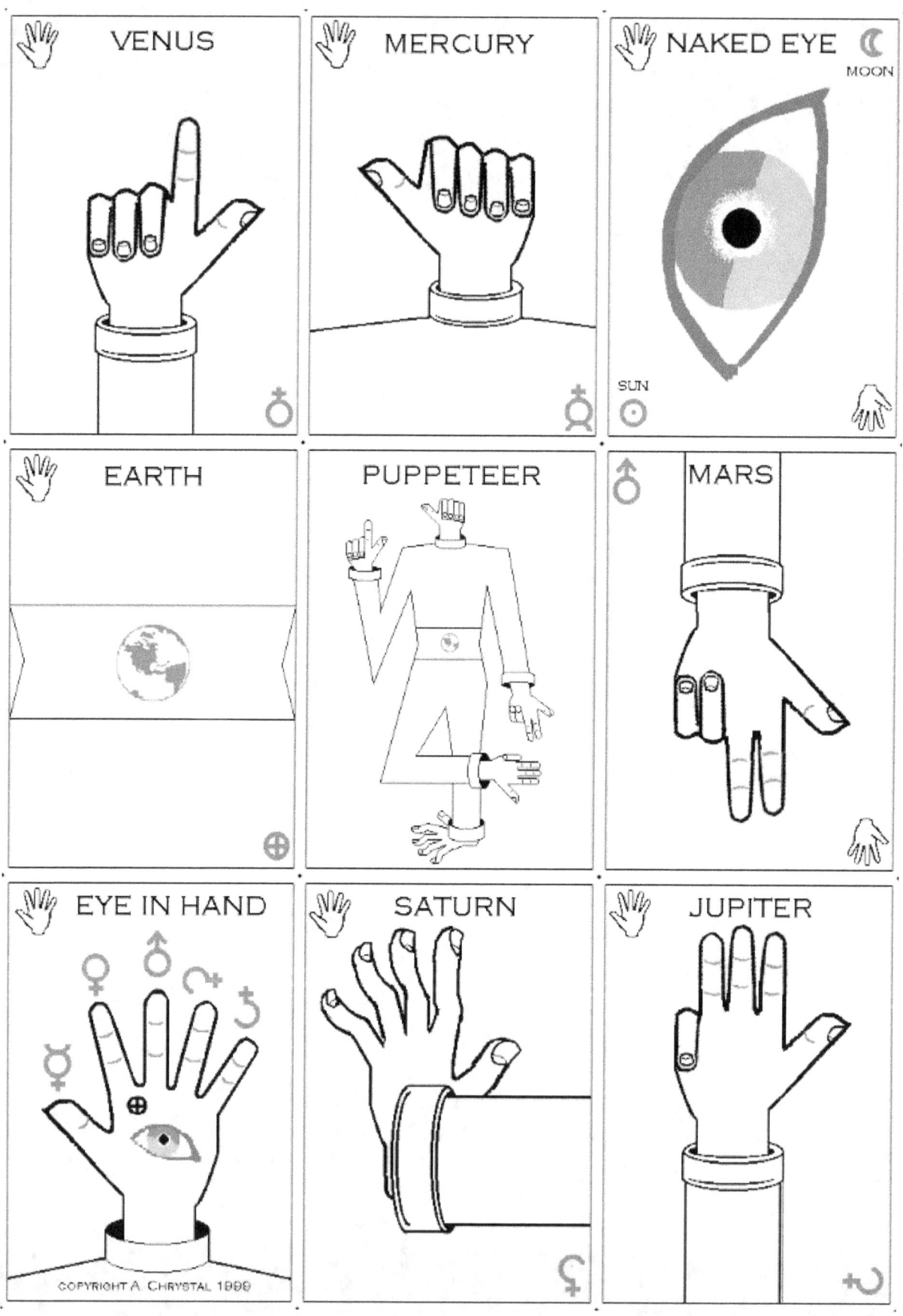

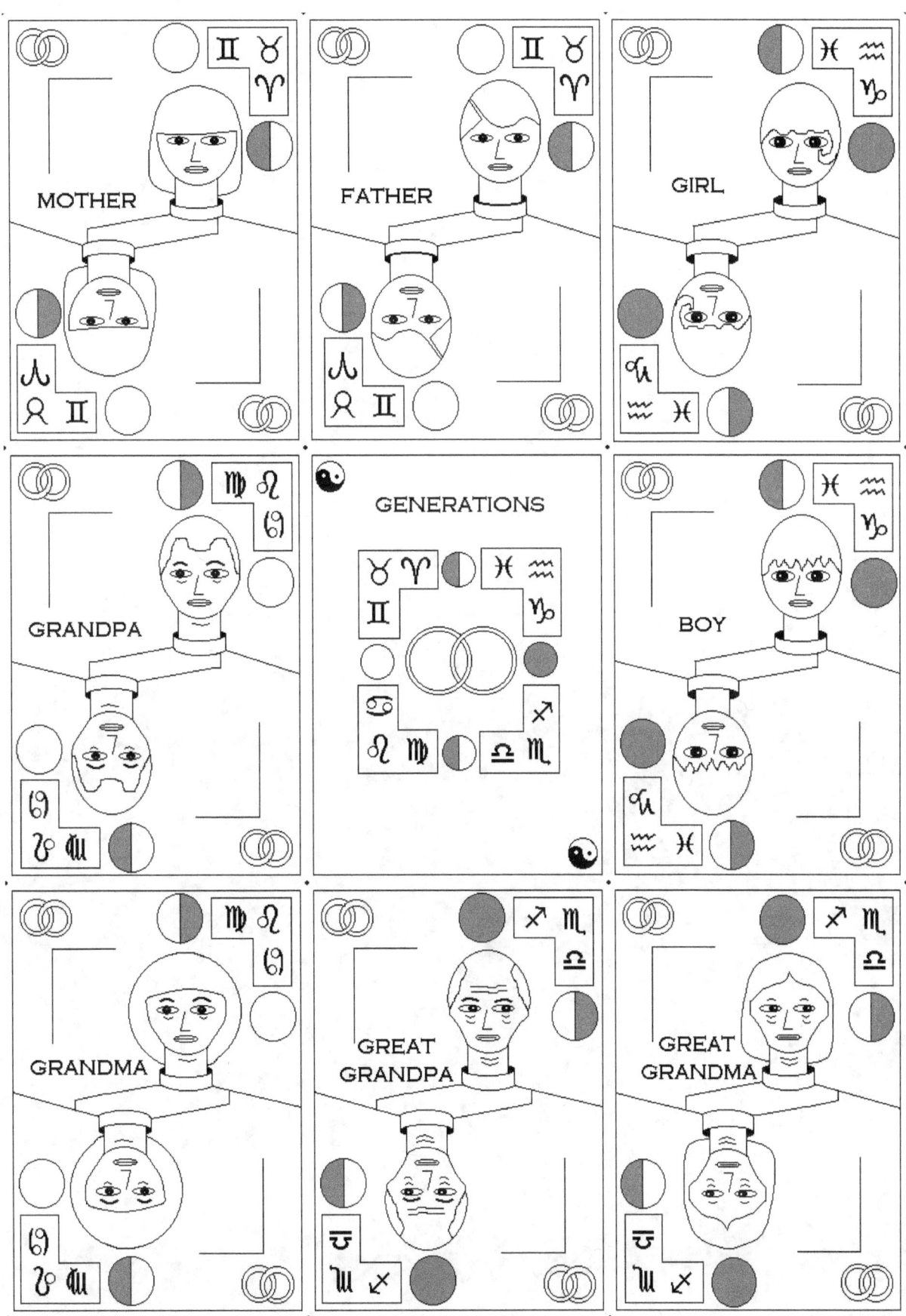

MOTHER

FATHER

GIRL

GRANDPA

GENERATIONS

BOY

GRANDMA

GREAT
GRANDPA

GREAT
GRANDMA

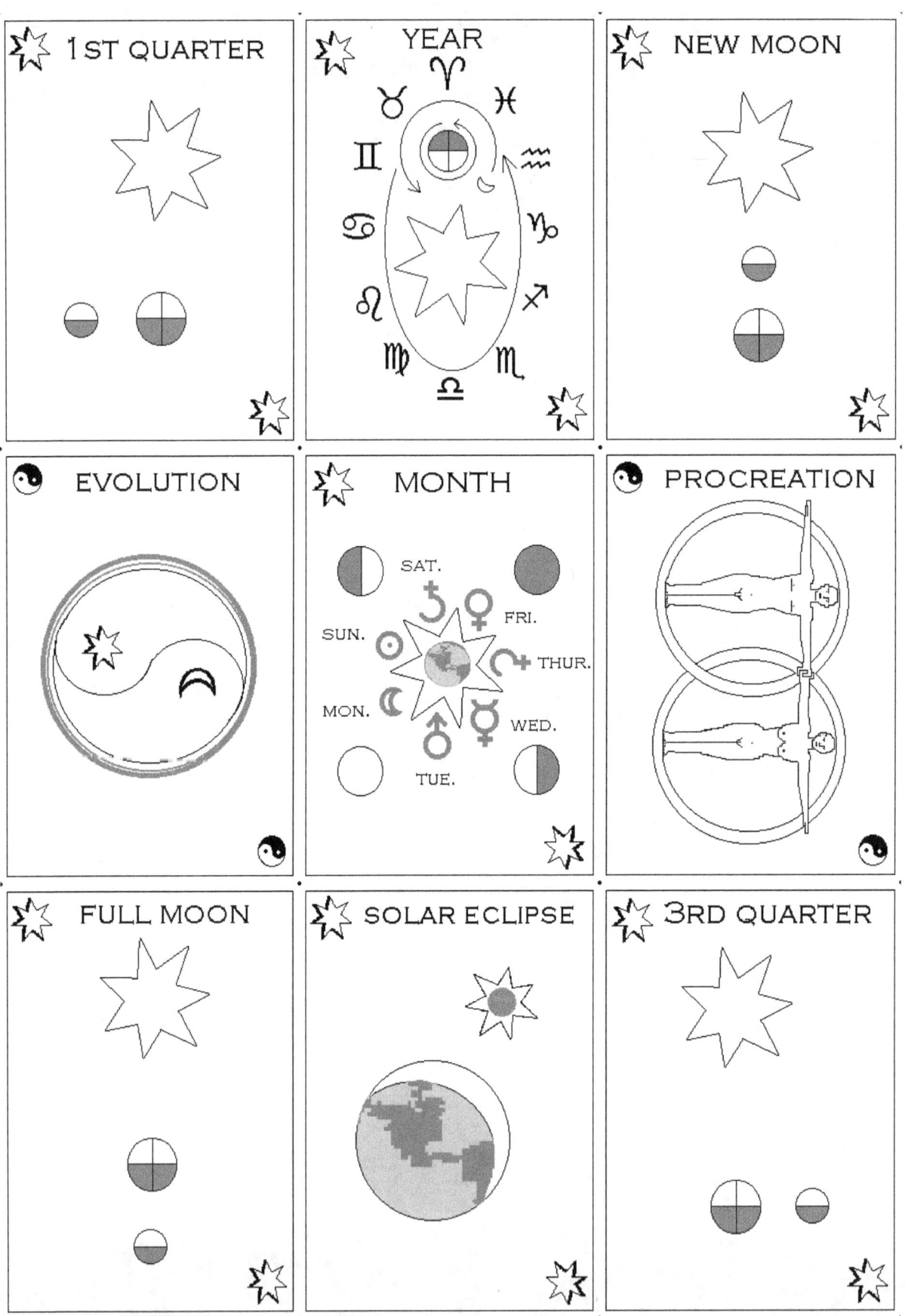

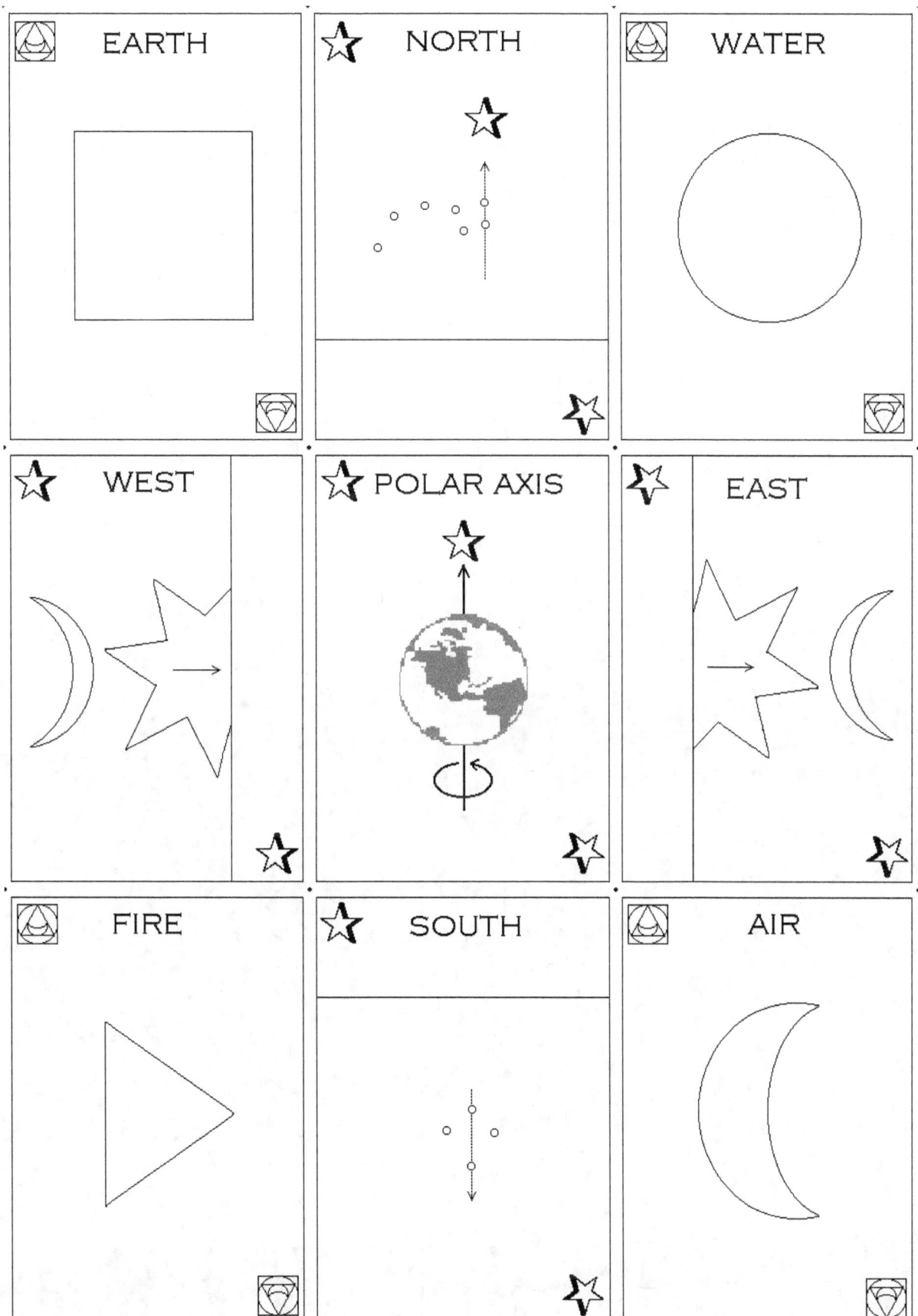

EARTH

NORTH

WATER

WEST

POLAR AXIS

EAST

FIRE

SOUTH

AIR

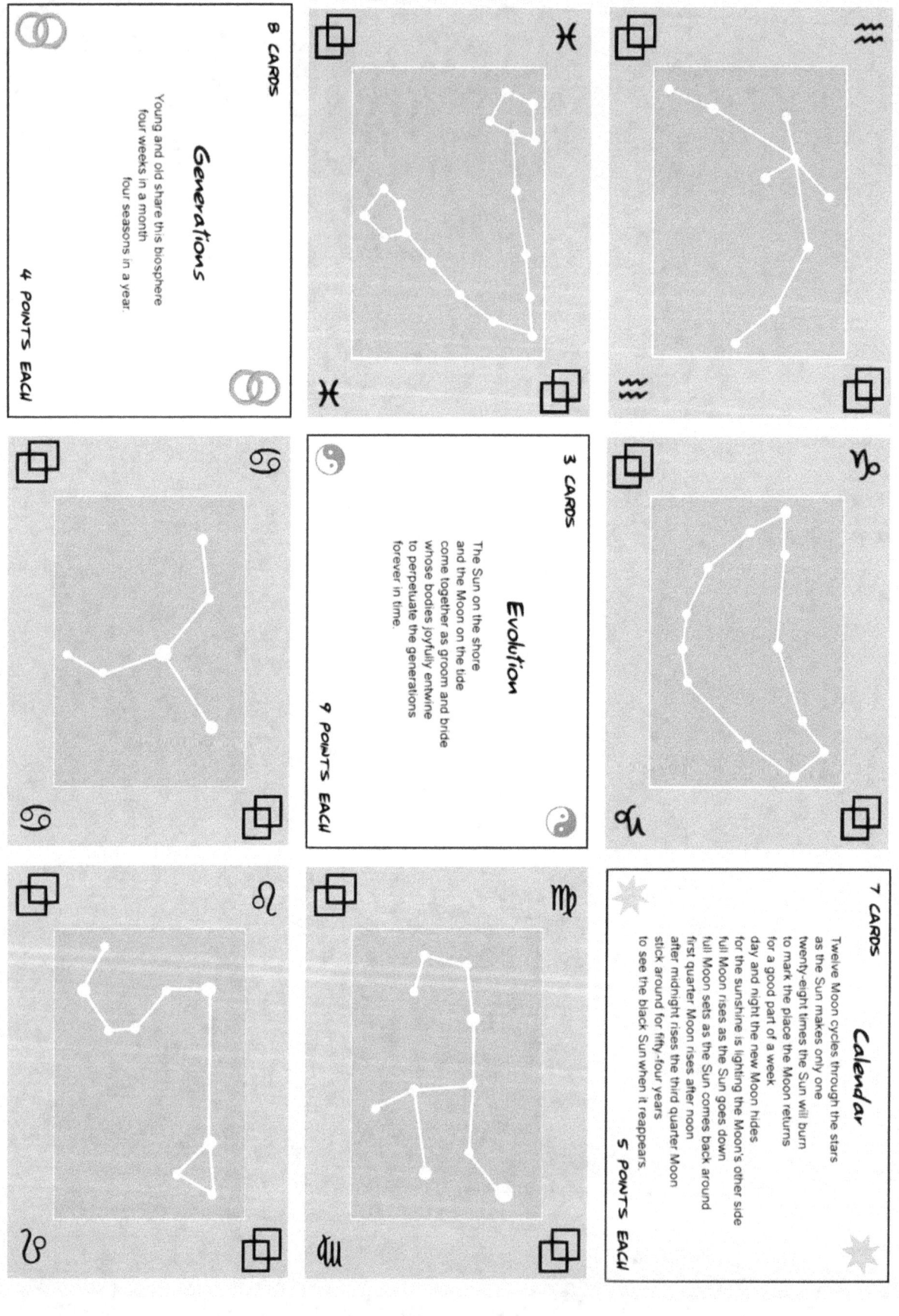

B CARDS

Generations

Young and old share this biosphere
four weeks in a month
four seasons in a year.

4 POINTS EACH

3 CARDS

Evolution

The Sun on the shore
and the Moon on the tide
come together as groom and bride
whose bodies joyfully entwine
to perpetuate the generations
forever in time.

9 POINTS EACH

7 CARDS

Calendar

Twelve Moon cycles through the stars
as the Sun makes only one
twenty-eight times the Sun will burn
to mark the place the Moon returns
for a good part of a week
day and night the new Moon hides
for the sunshine is lighting the Moon's other side
full Moon rises as the Sun goes down
first quarter Moon sets as the Sun comes back around
full Moon sets as the Sun comes back around
after midnight rises the third quarter Moon
stick around for fifty-four years
to see the black Sun when it reappears.

5 POINTS EACH

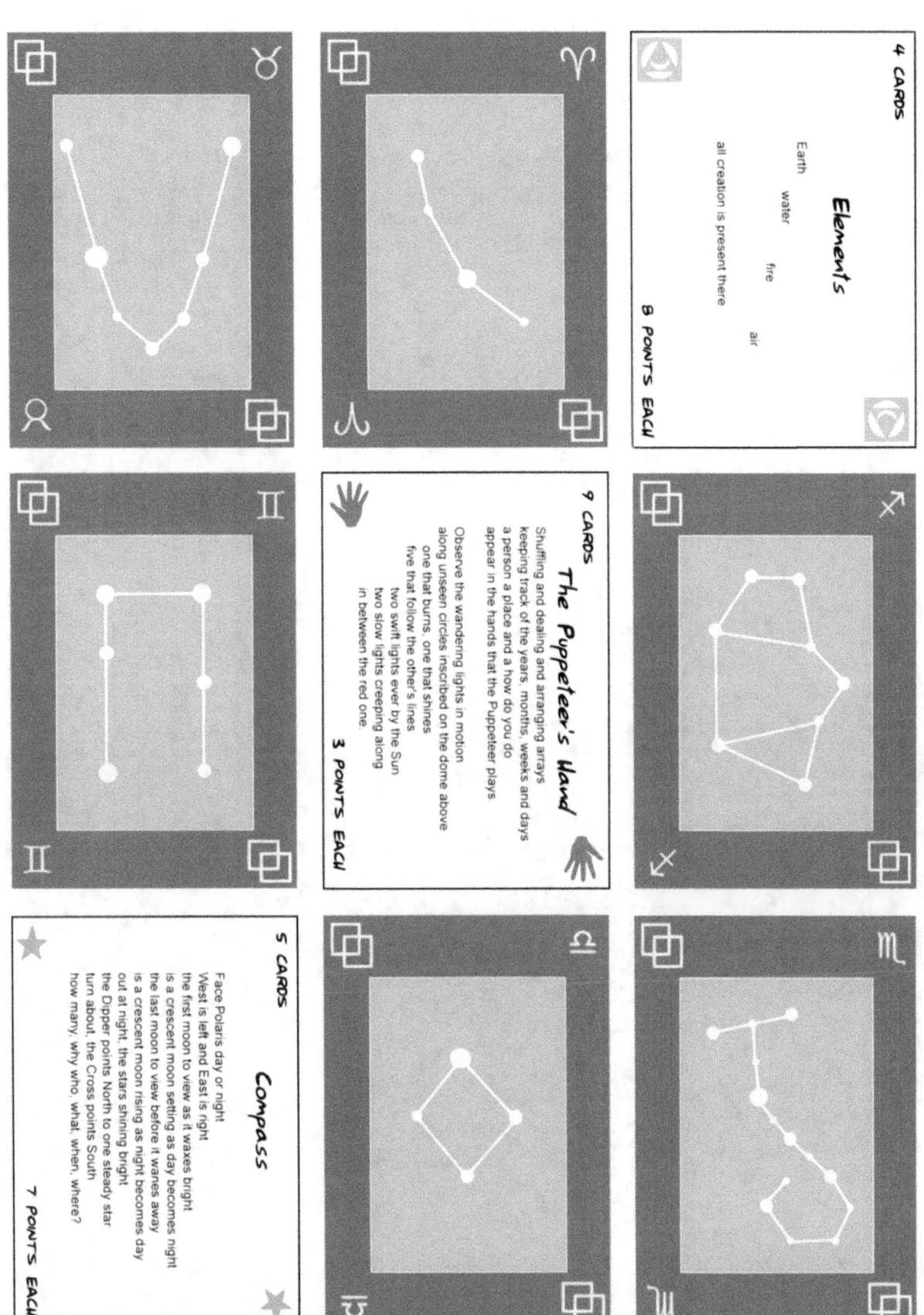

4 CARDS

Elements

Earth   water   fire   air

all creation is present there

8 POINTS EACH

---

9 CARDS

The Puppeteer's Hand

Shuffling and dealing and arranging arrays
keeping track of the years, months, weeks and days
a person a place and a how do you do
appear in the hands that the Puppeteer plays

Observe the wandering lights in motion
along unseen circles inscribed on the dome above
one that burns, one that shines
five that follow the other's lines
two swift lights ever by the Sun
two slow lights creeping along
in between the red one.

3 POINTS EACH

---

5 CARDS

Compass

Face Polaris day or night
West is left and East is right
the first moon to view as it waxes bright
is a crescent moon setting as day becomes night
the last moon to view before it wanes away
is a crescent moon rising as night becomes day
out at night, the stars shining bright
the Dipper points North to one steady star
turn about, the Cross points South
how many, why who, what, when, where?

7 POINTS EACH

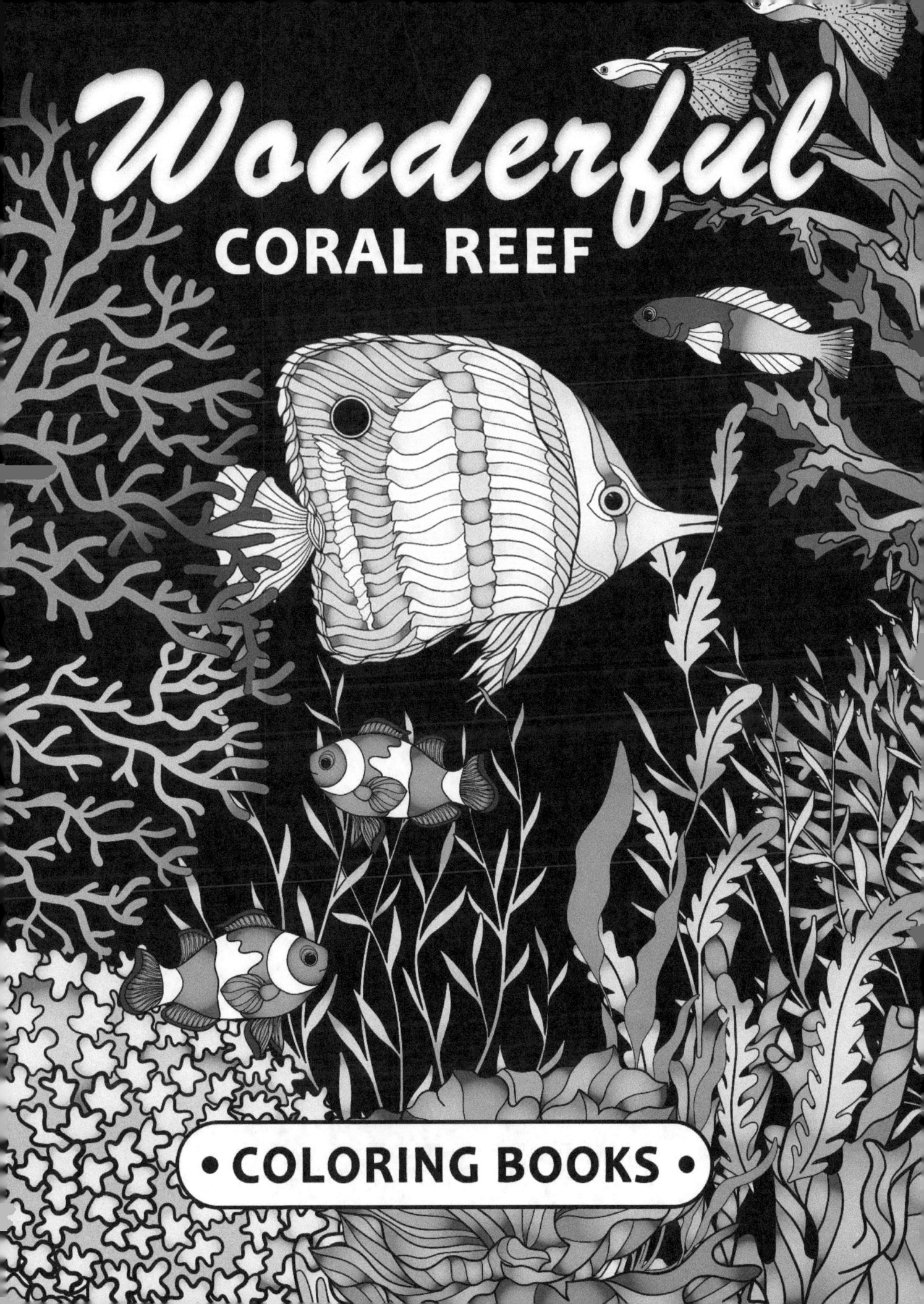

PUBLISHED IN 2018 BY
KODOMO PUBLISHING

PRINTED IN THE UNITED STATES OF AMERICA